MW01700718

Back to Basics:

How to create good jobs in the Pacific Northwest

by

Martin Jack Desmond

Forestry Financial Services, Inc.

Eugene, Oregon

Back to Basics: How to create good jobs in the Pacific Northwest
By Martin Jack Desmond

Published by:
Forestry Financial Services, Inc.
303 South 72nd Street
Springfield, Oregon 97478
www.forestryfinancial.com
541-968-5143

Publisher's - Cataloging In Publication

Desmond, Martin Jack

A comprehensive guide to creating good jobs in the Pacific Northwest by utilizing forests. /Martin Jack Desmond, with more than 60 charts and illustrations. 1st edition

Includes bibliograhical references and index.

ISBN 0-9758538-0-5

Library of Congress Control Number 2004094559

1. Business - Jobs
2. Environment - Forests
3. Economy - Forest industry

About the Author

Martin Jack Desmond has been a resident of the Pacific Northwest for over 25 years. His varied work and educational background provides him with a unique perspective on how to solve the economic and environmental problems facing the region.

Martin worked for the Forest Service and National Park Service on hot shot fire crews in the 1970s. He worked for five years for the Oregon Wilderness Coalition in the early 1980s and helped to pass the Oregon Wilderness Act which created nearly one million acres of wilderness.

He worked as a business manager for a large forestry contractor for eight years and worked as the executive director for the Northwest Forestry Contractors Association for seven years in the late 1980s and 1990s. He returned to college in 2002 and graduated with a Master of Accounting degree from the University of Oregon in June 2003. Martin passed the CPA exam in May 2003.

Northwest Forestry Contractors Association

The 501(c)(6) association was formed in 1989 by a group of licensed forestry contractors who were concerned about the non-compliance problems of the industry. The purpose of the association is to "level out the playing field" so that all contractors have the opportunity to compete fairly in business.

Acknowledgements

I would like to thank the Northwest Forestry Contractors Association Board of Directors for their support. Particular thanks goes to Dan Robertson, Dora and Pedro Zaldivar, and Tim Sweet for their support for the development of this book.

I would like to thank Jeff Peace of *Timber West Magazine* for putting on a great small log conference in Coeur d'Alene, Idaho that introduced me to new economic ideas. A number of Forest Service people deserve thanks for providing information. I would like to thank Jesse Springer of Springer Design for the front and back cover designs. I would like to thank Charles Spencer for content editing and Rachel Homrichhausen for copy editing. I would like to thank Wade Tullar for building a great Web site for me. I, however, am responsible for errors in the book.

Dedication

I would like to thank my partner, Vicki Megowan, and my father for their unwavering support.

Table of Contents

Table of Contents

Preface

Why are we so confused?

Imagine Henry Stamper, the tough old logger in Ken Kesey's *Sometimes A Great Notion* novel, traveling around the Pacific Northwest today. During his time, his logging friends had to struggle and curse against the mighty Wakonda Pacific timber company. Henry would likely be perplexed. He would wonder why the citizens of the Pacific Northwest had turned their backs on the most valuable natural resource of the region. Henry would ask why people toil at minimum wage jobs to attract flighty tourists from California when there is so much work and money to be made from the natural resources found in our forests.

This book lays out the vision of the Northwest Forestry Contractors Association of how we can create economically and environmentally sound jobs in the Pacific Northwest. Like the lost hiker who dies of starvation on a bed of miner's lettuce, we've turned our backs on our most important natural resource of the region. Our public and private forests are increasingly choked and dying from too much disease, growth, and fires. We've created a political environment where a fire burns 500,000 acres of the Kalmiopsis wilderness area. We then argue among ourselves of whether to harvest 50,000 acres of dead trees while allowing the gorse, tanoak, ceonothus, manzanita and other brush species to take over where once stood the ancient Kalmiopsis forest.

In the face of a stagnating Pacific Northwest economy, this book will argue that we can create tens of thousands of jobs that will rejuvenate forest lands and communities around the region. This book will discuss how we can create jobs that are more insulated from the ravages of global outsourcing that are so decimating our country. This book will describe how our region is an important carbon sink in the world and how we can play an important economic and ecological role in the storage of carbon dioxide in the face of accelerating global warming. Rather than inefficiently spending billions of dollars fighting wildfires, this book will describe how we can pay for the fuels reduction through timber sales and stewardship contracts. This book will discuss the potential for collecting biomass for generating ethanol to power our cars and reduce our dependence on oil imports from the Middle East. This book will discuss the constraints and limitations to job growth because of a lack of capital, compliance issues, stable and predictable flows of timber, and unwillingness to work together.

Mark Twain is reputed to have once said, "Everyone is always talking about the weather, but nobody is ever doing anything about it." We need lots of people to help step in and generate those jobs. We are like the encrusted Rohan's King Theoden in the *Lord of the Rings – Twin Towers*. We've become so mentally encrusted that we've become dull-witted and confused. Gandalf had to swing his walking stick and knock the king in the head to wake him up from the stupor. Without sounding too alarmist, our region and our country are faced with a stark choice.

If you don't have time to read this book, you can review Six Points to Economic Security. The six steps summarize what is said in the next 175 pages. The devil is in the details.

Six Points to Economic Security

Point 1) The United States and the Pacific Northwest have a "lack of good jobs" economic recovery.

Point 2) Global outsourcing is one of the factors for the loss of good jobs. America is unable to create jobs that are able to compete against low wages from Asia.

Point 3) Over three quarters of Forest Service lands in Oregon and Washington are at moderate to high risk from destructive wildfires because of large accumulations of brush and small trees.

Point 4) The federal government spent over $1.5 billion in both 2000 and 2002 to suppress wildfires.

Point 5) Rather than spend increasing sums of money on wildfire suppression, the federal government should sell small trees on our public lands to reduce the fuels accumulation. The sale of trees would generate funds to pay for more fuels reduction work.

Point 6) Companies and government agencies would hire Americans to work to reduce the fuels buildup and process the wood through mills. We could produce ethanol to power our vehicles and reduce our 60% dependence on Middle East and other foreign oil imports. We could produce more lumber to reduce our dependence for 30% of our supply from other countries. Wildfires would be eventually reduced. We could build up forests to store excess carbon dioxide to combat global warming.

Chapter 1

Our Current Economic Malaise

My year as a statistic ended last week. For 52 weeks, I've drawn unemployment insurance benefits, thereby being counted among the nation's millions of jobless persons. I remain jobless, but I was dropped from among those counted as being unemployed last week. I now join millions of others lost in the "economic recovery."

Former Associate dean of teacher education at Northwest Christian College
Opinion Piece, February 2004
The Register-Guard, Eugene, Oregon

Where, oh where, are the good jobs? It may be hard to remember the 1990s when the economy of the United States created 25 million new jobs as the unemployment rate dropped below 4% nationwide. By March 2004, the United States had lost a net 2.2 million jobs since the collapse of the dot-com bubble in 2001.

In this chapter

This chapter will discuss the loss of jobs, particularly good jobs, from the nation's point of view to provide a context for the regional problems faced in the Pacific Northwest. This chapter will discuss the threat of global outsourcing and loss of manufacturing jobs.

The US loss of good jobs is much worse

The actual job situation is much worse than the net loss of 2.2 million jobs in March 2004 (Figure 1-1). With economists estimating that the country needs to produce 150,000 new jobs per month just to stand still with new incoming workers, one would think that the unemployment rate would have risen even more. In addition to the loss of 2.2 million jobs since 2001, the country should have added another 5.4 million jobs in three years to stay even.

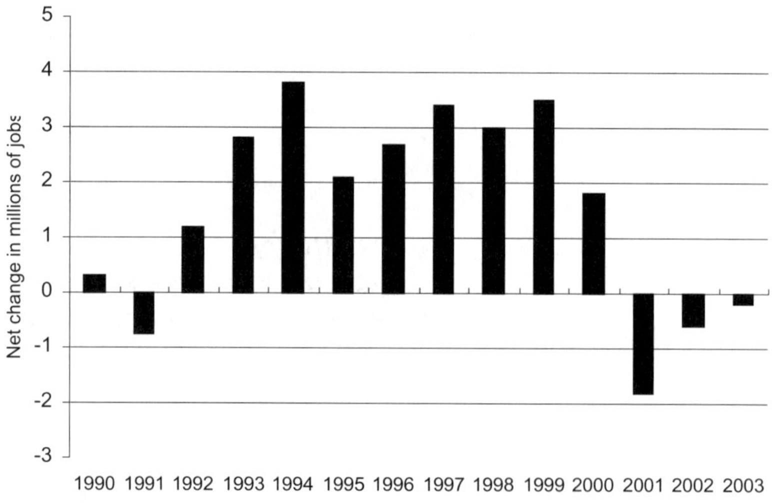

Figure 1.1 Loss of American jobs
Source: Bureau of Labor Statistics

The Labor Department reported that only 21,000 jobs were created nationwide in February 2004. Yet, the unemployment rate stood still at 5.6% across the nation. According to Scott Nelson, reporter for NPR, the main reason that the unemployment rate stood still "was because so many people left the work force, 400,000 in February alone."

Americans are worried. A winter 2004 Gallup poll found that 41% of Americans view jobs or the sluggish economy as their personal top concern – three times the number for either the Iraq war, exploding health costs, or terrorism. In the spring of 2004, job creation did begin to improve with the creation of hundreds of thousands of jobs around the country. Even the unemployment rate in Oregon dropped below 7%.

Are all developed countries losing good jobs?

The United States gross domestic product grew by a blistering 8.2 per cent in the third quarter 2003, but good job creation has been stalling ever since the economy was reported to be climbing out of the recession. Despite having an economy that is one-eleventh the size of the United States, Canada generated 53,000 new jobs in December 2003 while the United States created only 1,000 new jobs.

What is happening to the American job machine? According to Mr. David Wyse, Chief Economist for Standard & Poor's, "There's a fundamental change occurring in the way American corporations are organized. They don't want employees anymore. We've gone from 10 years of moving to just-in-time inventories to now moving to just-in-time employment."

According to Georgetown University economics professor Harry Holzer, the United States has not suffered such a prolonged decline in jobs since the 1930s Depression. Holzer said that 4 million American workers have run through their unemployment benefits without finding jobs.

Compounding the lack of good jobs creation are the twin evils of federal and trade deficits looming just off into the future. The federal deficit hit $380 billion in fiscal year 2003 and is expected to cross over the $445 billion threshold in FY 2004. American consumers continue to purchase low-cost goods from across the world pushing the annual trade deficit over $500 billion.

If job and economic conditions are described as "difficult" in other parts of the country, the job situation in Oregon and

Washington could be better characterized as "terrifying." The states of Oregon and Washington lost tens of thousands of manufacturing jobs in three years with little expectation to recover any of those manufacturing jobs.

Global outsourcing pushes down wages

The plight of blue collar workers in the United States has been well-known for some time. The manufacturing sector has been steadily losing jobs for four years without any sign of slowing down. More recently, white collar workers are suffering the same problem.

I was surfing on Google one night and typed in the keywords "bookkeeping services." I clicked on one sponsored link and was immediately transferred to a site in India. The company offered CPA-equivalent employees for $5.95 per hour if you signed up for a long term package. Otherwise, the charge was $8 per hour for short-term use of their CPAs.

Forrester Research released a study showing that 3.3 million white collar jobs would be moving offshore by 2015. The Gartner research firm projected that as many as 5% of all information technology jobs could be moving abroad between mid-2003 and the end of 2004. While many white collar workers were willing to tolerate the loss of blue collar jobs under the belief that the United States was moving to an information technology economy in which many Americans would benefit, the specter of so many white collar jobs moving offshore is unsettling. A University of Maryland poll conducted in the winter 2004 found that only 28% of those earning more than $100,000 supported active free trade – down from 57% in 1999

The rise of the Internet, the low cost of bandwidth and the lure of lower offshore wage costs are much too irresistible to multi-national corporations. One columnist noted that there are almost three billion people in China, Russia, and India. If one simply dismisses 90% of those people as illiterate peasants, there would still be 300 million increasingly well-educated and increasingly English-savvy workers who will be able to undercut the wages and productivity of Americans. The bottom-line is that Americans will not be able to compete against rock-bottom wages – no matter how efficient and productive. Americans need to be performing types of work that are not able to be replicated in China, India, Vietnam, Indonesia, Thailand, Bangladesh, or a host of other developing countries.

Economists don't understand the real threats

Economists – free from the desperate worry of trying to find a job like many Americans – are rather sanguine about the deleterious effects of job outsourcing.

I spent 1 ½ years in mid-life to go to the Lundquist College of Business at the University of Oregon with the idea that there would be a good-paying job waiting for me when I graduated. While researching for this book, I found a memorable quote from my dean at the Lundquist College of Business. *Oregonian* news reporter Gail Kinsley Hill asked the dean about the controversy over job outsourcing in a January 18th, 2004 article.

The dean replied that he found "benefits in the aggregate." "There's no doubt this off-shoring phenomenon is, on net, good for the economy," he said. He said developing countries benefit from more jobs, U.S. consumers enjoy lower-priced goods and services, and U.S. companies are able to cut costs.

The dean replied that he was not concerned about replacement jobs, provided that there was innovation and entrepreneurship. Which jobs, in particular, will crop up? That "I don't know," he said. "If I did, I'd be a venture capitalist."

If global outsourcing is good for our country, it occured to me that the UO business school could beam in video and audio presentations over the broadband Internet from Bangalore, India accounting professors at $8 per hour and fire the existing American professors. UO students could email their homework and tests to the Bangalore, India professors. Everyone would benefit because tuition costs would be reduced.

The jobs threat posed by globalization is very real – particularly for the Pacific Northwest because of its geographical proximity to Asia. In order for the Pacific Northwest to survive and prosper without the threat of a continued downward spiral, the region must create good jobs that are unique to this area.

Oregon and Washington used to have many types of manufacturing jobs that took advantage of the natural resources of the region. One such manufacturing industry was aluminum. Aluminum requires lots of electricity. Aluminum manufacturers were able to take advantage of the low electricity costs of the Pacific Northwest because of the power-generating dams along the Columbia River. As electric rates rose and supply diminished, the aluminum manufacturers disappeared one by one in the 1990s.

Quickie

By March 2004, the US had lost 2.2 million net jobs since 2001 The 2.2 million net loss does not include approximately 5.4 million jobs that needed to be generated during this period of time to stay even with the monthly need to create 150,000 jobs. Global outsourcing is projected to move 3.3 million white-collar jobs offshore by 2015.

Chapter 2

What's happened to our good jobs in the Pacific Northwest?

Even today, the Northwest Forest Plan remains controversial. Some feel that it produces timber harvests that are too small, resulting in economic and social harm to the region. Others feel that timber harvests are too large, threatening the survival of plant and animal species . . . However, based on three accomplishments alone, the NW Forest Plan must be judged as one of the singular achievements of the Clinton Administration in the natural resources field.

James Pipken, Director of the Office of Policy Analysis, USDA, Northwest Forest Plan Revisited 1998

The last five years have been tougher on the Pacific Northwest than most other regions of the country. Oregon has single-handedly led the nation with the highest unemployment rate for much of the last three years. The state of Washington was not far behind. Both states have had total unemployment rates from 1.7 to 1.9 percent points higher than the rest of the nation (Figure 2.1). The citizens of both states have failed to step forward to develop a cohesive economic strategy to help lift the hopes of everyone in the region.

In this chapter

This chapter will discuss what has happened to the job markets of Oregon and Washington during the past decade. The chapter will include a big picture view of the types of jobs in the Pacific Northwest. The chapter will focus upon the impacts to forest workers in this region in the 1990s. Our Pacific Northwest region has allowed hundreds of thousands of hard working people to lose their jobs and careers.

Laissez faire attitude of our citizens

The citizens of both states had a somewhat laissez faire attitude about the wood products industry in the 1990s. The feeling of many hot-shot economists was that the wood products industry was a smokestack industry that needed to be replaced by human beings working in front of humming Dell or Gateway networked computers. Everyone

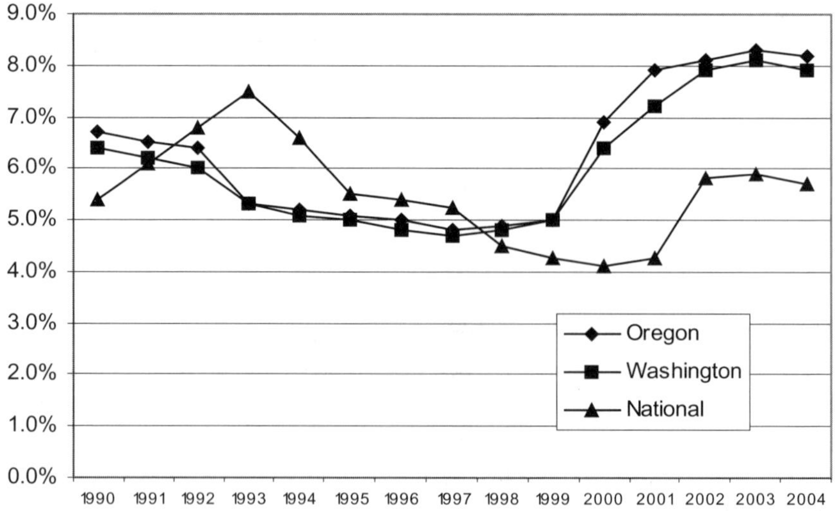

Figure 2.1 US, Washington, and Oregon unemployment rates

Source: Oregon Employment Department, Washington Employment Security Department would become rich writing HTML code to produce web pages. Even I was building Web pages for companies in the 1990s. Enormous numbers of high paying information jobs would be created along the Digital Highway. The NASDAQ roared up to the 5,000 mark with certain economic pundits predicting that the DOW would hit 30,000.

In Oregon, the Tillamook State Forest would be morphed into the Silicon Forest as workers migrated from cutting trees to cutting and pasting on computer screens.

One day in 2001, people realized that the Emperor wasn't wearing any clothes. The NASDAQ began its heady descent back to Earth and the DOW was not too far behind.

The loss of manufacturing jobs

The US has lost 2.8 million factory jobs in the last several years. Washington has lost a staggering 46,600 manufacturing jobs since 2001. Since Washington's job peak in 1998, Washington has seen the total loss of 94,400 jobs. Washington's aerospace industry is very volatile, accounting for 16% of the annual change in total employment in Washington since 1990. Washington's manufacturing jobs averaged 200,000 jobs in the 1940's, 1950's, and 1960's, reflecting nearly 35% of the total number of jobs in the state. Today, there are 350,000 manufacturing jobs, but these manufacturing jobs represent less than 15% of the total number of jobs.

The US Bureau of Labor and Statistics requires states to organize employment data in broad major industrial divisions. This standardized approach allows economists to compare the relative strengths of each state.

The average annual wage of workers in Washington and Oregon in 2002 was $38,244 and $33,684 respectively (Table 2-1 and Table 2-2). The average annual wage is somewhat inaccurate. That figure is deceptive since the average hourly wage includes individuals who are making hundreds of thousands of dollars. The average hourly wage in Oregon in 2002 was $17.56 per hour. A more accurate assessment is the median hourly wage figure which was $12 in 2002. The median represents the halfway point in the sample size. Or, in simple English, half of all Oregonians earned less than $12 per hour and one half earned more than $12 per hour.

Table 2-1: Washington Employment and Wages - 2002

Major Industry Division	Total wages (billions $)	Employment	Annual wage
Trade, transportation, utilities	$16.3	494,272	$33,087
Government	$19.3	490,324	$39,360
Education and health services	$9.1	287,410	$31,725
Manufacturing	$14.3	280,964	$50,901
Professional and business services	$12.9	279,997	$45,933
Leisure and Hospitality	$3.7	240,611	$15,297
Construction	$5.6	142,885	$39,396
Financial activities	$6.5	141,710	$45,883
Information	$9.5	92,715	$102,309
Forestry, Fishing, Mining, Agriculture	$1.7	79,354	$20,963
Other services	$1.9	74,461	$25,336
Total	$101.1	2,643,715	$38,244

Source: Washington Employment Security Department

For example, the average wage of the information worker in Washington is $102,309 for 2002 (Table 2-1). Table 2-2

Oregon Employment and Wages - 2002 shows a more down-to-earth wage of $47,059 for information workers. One wonders if the average wage for Washington information workers is over twice the Oregon average wage because of the existence of so many stock option millionaires who work for Microsoft in Redmond, Washington.

Table 2-2: Oregon Employment and Wages - 2002

Major Industry Division	Total wages (billions $)	Employment	Annual wage
Trade, transportation, utilities	$9.8	313,722	$31,284
Government	$9.3	252,932	$36,913
Education and health services	$5.8	179,417	$32,486
Manufacturing	$8.7	201,222	$43,600
Professional and business services	$6.4	172,384	$37,418
Leisure and Hospitality	$2.0	149,283	$14,062
Construction	$3.0	142,885	$39,396
Financial activities	$6.5	77,909	$39,483
Information	$1.7	36,211	$47,059
Forestry, Fishing, Mining, Agriculture	$1.1	46,819	$23,641
Other services	$1.3	58,093	$22,615
Total	$52.9	1,573,083	$33,684

Source: Oregon Employment Department

Wages shown in Table 2-1 and 2-2 do not include total personal income. Total personal income includes other forms such as capital gains, interest, rental income, etc. Washington leads the Northwest in total personal income at $198 billion in 2002 (Table 2-3). Although Washington's and Oregon's total personal income was far greater than the other Northwest states, other states outpaced both states in terms of real growth from 2001 to 2002. Alaska grew 2.7% in personal income (Table 2-3).

Table 2-3: Total Personal Income for Northwest States and the United States (billions of dollars)

Area	2002	Real change
Washington	$194.2	2.0%
Oregon	$100.4	1.1%
Idaho	$33.5	2.4%
Montana	$22.6	2.6%
Alaska	$20.4	2.7%
United States	$8,891	1.1%

Source: Washington State Employment Security Dept.

Service industry jobs pay less than manufacturing jobs

Manufacturing jobs are important because these jobs form the base of the economy. New wealth is created from manufacturing jobs while most service jobs simply circulate existing money.

Many economists used to like to talk about Information Age jobs. Information Age jobs, when available, do pay quite well. The problem is that Information Age jobs are not available to the typical resident of Oregon or Washington who do not have advanced degrees or advanced programming and computer skills. Rather than focus on high tech jobs that are easily replicated in Bangalore, India or other places in the Far East, we should focus on manufacturing jobs to utilize the unique natural resources of the region.

Manufacturing jobs pay far more than Oregon and Washington waitresses and waiters serving food to California tourists. The average manufacturing job in Washington pays slightly more than $50,000 while service industry jobs pay an average of $15,300 (Table 2-1). In Oregon, the spread is similar. The Leisure and Hospitality Service industry pays an average of $14,062 while the Manufacturing industry pays $43,600 (Table 2-2).

New jobs pay less than existing jobs

One very serious problem facing any new job creation is the significant trending down of wages. A national survey in spring 2004 showed that industries adding jobs such as retail and health care pay 21% less than factory jobs. The more extreme examples of wage decline have occurred in states such as Massachusetts where new jobs pay 40% less than manufacturing jobs. Wage decline has averaged 29% less for new Ohio workers, 26% less for new Michigan workers, and 23% less for new Pennsylvania workers.

Seattle - Washington's epicenter of the 2001 recession

Yet another problem facing the Pacific Northwest is the significant loss of jobs that occurred in the largest metropolitan areas of the region. Typically, the Portland and Seattle metropolitan areas had lower rates of unemployment and more job creation than the rest of the region during most of the 1990s.

However, the Seattle Metropolitan area suffered the biggest brunt of the 2001 recession. The Seattle Metropolitan area lost 94,700 jobs from September 1999 to September 2003 while the rest of the state actually gained 36,000 jobs. The job losses in the Seattle metropolitan area dominate the rest of the state as the area accounts for one half of the state's employment.

That's the big picture of the types of jobs and wages paid to workers in the Pacific Northwest. The chapter will now focus upon the impacts for forest workers. Many of these forestry workers did not enjoy the fruits of the 1990s because they were in a difficult period of transition.

What happened to the forest workers during the 1990s?

As the likelihood appeared of a collapse of the federal timber sale program in the early 1990s, timber industry and environmental pundits on both sides were claiming that the other side was exaggerating or misleading the public about the potential job losses.

Oregonian reporter Pete Sleeth wrote a May 21, 1995 article entitled "*How many jobs did you say?*". Sleeth wrote, "George Bush and the timber industry were wrong. For that matter, so were the environmentalists." Sleeth's data was based upon a review of job data from the state employment departments of Oregon, Washington, and Idaho. Sleeth stated that 16,695 direct wood products jobs were lost in Oregon, Washington, and Idaho, much less than the 65,700 projected by the Northwest Forest Resource Council in 1989. Sleeth also chided environmentalists for "underestimating the economic impact or glossing over the job losses as a natural function of a dying industry."

According to Paul Ehinger, a forestry consultant who tracks mill closures, 251 sawmills have closed affecting approximately 25,000 workers in the past decade.

Getting another job isn't easy for a millworker

Timely snapshots are useful, but may leave out some long-term trends since all of the data is not yet available. One study, entitled *Employment Transitions in Oregon's Wood Products Sector during the 1990s* (Journal of Forestry, June 2003), looked back at a key period of time from 1989 to 1991. Written by Darius Adams of Oregon State University, Ted Helvoigt of ECO Northwest, and Art Ayre of the Oregon Employment Department, the study analyzed 60,000 workers who had been defined as working in the wood products industrial code during the period from the fourth quarter 1989 to the third quarter 1991. Employers are required to report the wages of all workers covered by unemployment insurance. The Employment Department estimates that 88% of all workers were covered by unemployment insurance in 1990.

What happened to those 60,000 human beings – at least in terms of work by the date of the study in 1998? The study found that 25,000 (42 percent) were still working in the wood products industry. Another 18,000 (30 percent) were employed in another covered work activity in Oregon (Figure 2-2). The remaining 17,000 (28 percent) simply disappeared from the wage reports.

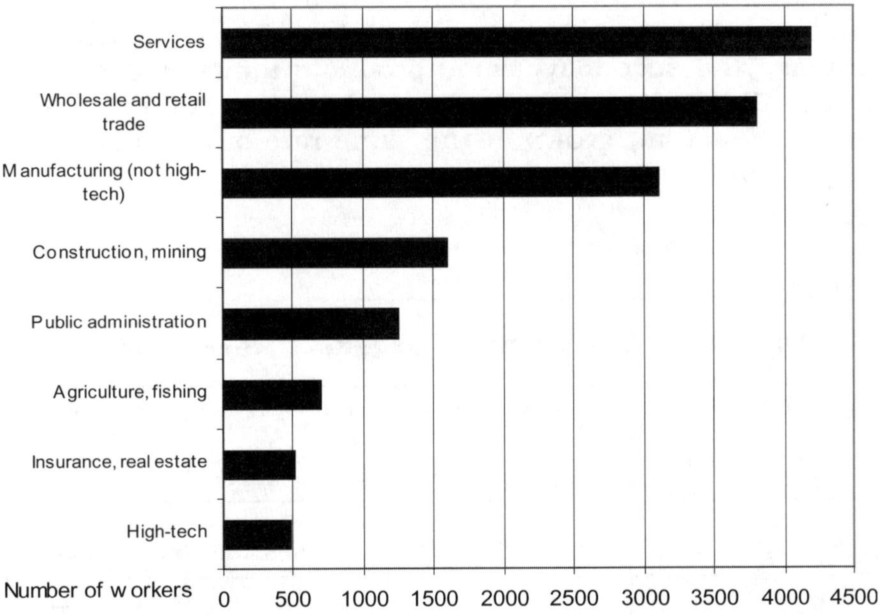

Figure 2.2: Employment of workers who left the wood products industry

Source: Employment Transitions in Oregon's Wood Products Sector during the 1990s, Journal of Forestry, June 2003.

Of the workers who stayed in the wood products industry, their mean annual wage increased from $30,280 to $32,120 in the eight year period. Of the workers who left the wood products industry, their mean annual wage declined from $21,450 to $21,260 over the eight year period.

The largest category of displaced workers who took jobs in the services industry saw their mean annual wage decline below $15,000 (Figure 2-2). Although the high tech industry was going to rescue these workers, only 500 of the 18,000 transitioned workers were able to find work in the high tech industry by 1998.

What is the current state of the wood products industry?

Oregon's primary and secondary wood products industry generates $10 billion in annual sales, providing over 60,000 direct employment jobs, $2.89 billion in workers' wages and $1.1 billion in timber sales today. Ten years ago, the primary and secondary wood products industry generated nearly that same amount of money so the wood products industry has not grown at the same rate of the rest of the sectors in the state of Oregon.

Table 2-4: Oregon wood products wages - 2002

Job type	Total wages	Employment	Annual wage
Logger	$ 293,897,301	7,984	$36,801
Agriculture and forestry support	$ 189,978,719	8,465	$22,422
Wood products manufacturing	$1,130,452,956	32,405	$34,855
Paper manufacturing	$ 396,551,956	6,911	$57,379
Furniture and related manufacturing	$ 217,713,841	7,449	$29,227
Source: Oregon Employment Department			

In Washington, the Forestry, Fishing, Mining, Natural Resources, and Agriculture major industrial division employed 79,354 workers in 2002 with total wages of $1.7 billion for an average annual wage of $20,963 (Table 2-1). This group is "ambiguous" because of the different types of work. For example, lumber mill workers are classified in the Manufacturing major industry division where the average wage is $50,901 (Table 2-1).

Statistics can be ruthlessly inaccurate

I have worked for over 15 years in the forestry services industry. Forestry services work tends to be seasonal in nature with workers transitioning in and out of the industry on a weekly and monthly basis. Whenever I would look at the 851 forestry services industry class wages and number of workers published by the Oregon Employment Department, I knew that the data understated the total number of workers. The forestry services industry includes an unknown percentage of workers who are paid cash under the table and never appear on any W-2 statements.

The forestry services industry has a large component of undocumented workers who are using illegal social security numbers. These workers will likely not get any credit for taxes paid into that social security number so these workers do not have any incentive to have taxes taken out of their wages. Furthermore, an employer who uses undocumented workers knows that they can avoid paying payroll taxes and workers compensation premiums. So, numbers are not always an accurate reflection of life.

If there are any conclusions to be drawn about the plight of forestry workers, the data suggest that it is not easy for a person in the mid-stream of their life to transfer to another occupation. The data suggests that mill workers who stayed in the wood products industry generally earned more money than other workers who tried to venture to other "greener" pastures.

Quickie

Oregon and Washington have lead much of the nation during the past three years with the highest unemployment rates in the country. Tens of thousands of manufacturing jobs have been lost in the last decade. The wood products industry's job loss has stabilized over the past several years. However, forestry workers have had difficulty transitioning into other industries.

Chapter 3

The Five Forces of Job Creation

There's no business like show business, but there are several businesses like accounting.

David Letterman

While I was attending the Lundquist College of Business at the University of Oregon in 2002, the accounting and finance professors were quite enamored with a professor who had attracted worldwide attention from the business community. Michael Porter, author of *Competitive Strategy: Techniques for Analyzing Industries and Competitors*, had been marketing a concept called the "Five Forces Model." The business community was particularly attracted to the Five Forces Model because the concept actually made good business sense unlike many other theoretical economic models by university professor.

In this chapter

It is important to understand how jobs can be created and maintained that will survive the rigors of the global marketplace. If Americans are going to maintain some semblance of their current living standards, they will need to figure out how to compete in a global market against hundreds of millions of people who will work for $.40 per hour.

How the Five Forces Model works

The Oregon Employment Department uses the term "industry clusters" as geographical groups of business linked together by product and supply chains. These businesses are related by common markets and technologies as well as by worker skills (www.qualityinfo.org/olmisj). Forest/wood/paper products are considered one of the 11 traded industry clusters in Oregon.

Michael Porter's Five Forces model, presented in *Competitive Strategy: Techniques for Analyzing Industries and Competitors*, offers a useful tool for understanding this industry cluster. (Figure 3-1).

Porter's theoretical model sheds light on the changing forestry jobs that were created in the Pacific Northwest in the last decade.

Degree of rivalry

Basic economics 101 class states that more competitors and rivals competing against each other will put a downward pressure on price increases and profit. Rivalry is influenced by many factors including:

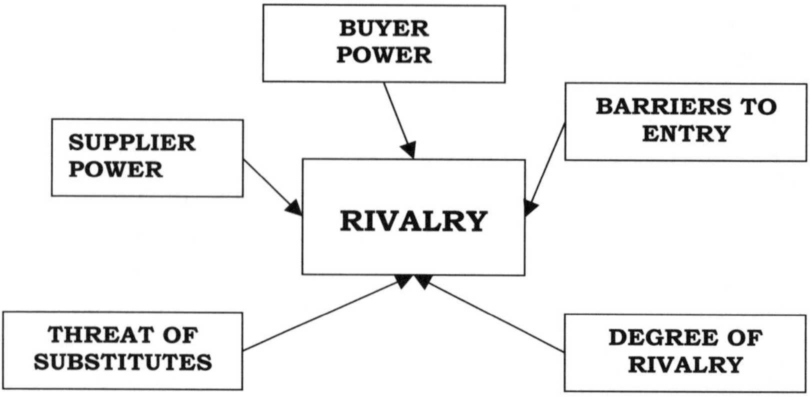

Figure 3-1: Five Forces Model

1) Number of firms in competition
2) Market growth
3) Product differentiation
4) Exit barriers
5) Diversity of rivals
6) Fixed costs
7) Switching costs

While some industries are characterized by a concentration of firms holding the majority of the market share, the paper and wood products industry is highly competitive. No one particular company holds a position in which a company could dictate prices nationally (Note: Weyerhaeuser has either lost in court or settled a number of cases involving monopolistic alder log buying activities in 2004).

While the timber and forestry services industry can be cutthroat in its competition, there is a surprisingly high level of coordination between different companies. Logs removed from Weyerhaeuser may be sent to Roseburg Forest Products from one unit while logs removed from a nearby Roseburg Forest Products unit may be sent to yet another mill that has better capacity to process the logs. The cooperative effort of the mills allows for better efficiency and utilization.

What puts Oregon and Washington in a good position to create lots of new forestry and logging jobs is the existence of a large number of inter-locking chip and sawmills able to bid for raw materials and able to process lumber.

Threat of substitutes:

Threat of substitutes refers to other products in other industries that could be substituted for wood. The wood products industry is faced with some types of threats of substitution, but most Americans prefer to live in wood houses and use paper products. By contrast, most commercial buildings are built with substitutes such as steel, cement, rock, metal or bricks.

Where the threat of substitutes affects the American wood and paper products industry is the import of Canadian lumber. Canadian imports amount to nearly 30% of the lumber consumed in the U.S.

There are no Douglas-fir trees growing in India or China. Unless we ship the Douglas-fir logs to India or China, we do not have to worry about the Chinese or Indians or Indonesians or Vietnamese manufacturing products at $.40 per hour. Douglas-fir and ponderosa pine trees are our trump card. Except for the Canadians, there are no other countries that have the capability to grow Douglas-fir and ponderosa pine. Of course, there are huge plantations of radiata pine growing in New Zealand and Chile that are owned, in part, by American forestry corporations.

Supplier power

Supplier power refers to the need of the buyer to purchase raw materials from the supplier. For example, supplier power is considered powerful if the suppliers are concentrated or if there is a significant cost to switch suppliers. Likewise, supplier power is considered weak if there are many competitive suppliers or if commodity products are standardized. Highly concentrated purchasers may weaken suppliers.

Buyer power

Buyer power can have a significant impact on an industry if buying is concentrated among a few purchasers. If a few buyers purchase a significant amount of the product, the producing industry is subject to what economists call a monopsony. A monopsony is a market in which there are many suppliers, but only a few buyers. As will be discussed later in the chapter, a monospony may have problems with not enough suppliers during peak periods.

Barriers to entry

Under a theoretical capitalistic system, firms would be free to enter and exit a market. Profits would be minimal since the firms would quickly move into other sectors of the business whenever there was a profit to be generated. However, there are a number of natural and unnatural barriers to prevent firms from actively moving quickly into another industry sector.

1) Government regulation of markets such as utilities
2) Patents and proprietary knowledge
3) Fixed assets such as equipment
4) Organizational economies of scale
5) Difficulty in switching brands
6) High exit costs

Are there any real world examples?

Here is a somewhat well-known fact among Oregonians. Oregon's fire contractors provide 90% of the private sector fire crews in the nation. With the exodus of National Guard and Army reserve troops to Iraq, Afghanistan, Haiti and other friendly places in world, the federal and state land management agencies are having to increasingly rely on private sector fire crews to deal with the explosive fires. How did a bunch of yahoo contractors in Oregon manage such a monopoly?

In the early 1990s, forestry service contractors recognized that they would have to evolve and transform their businesses. With the collapse of the federal timber sale program during the period of 1991 – 1993, the frog pond was going to shrink roughly in half for intensive forestry management work. Some frogs needed to evolve into toads or wart hogs if they wanted to survive.

Loggers, foresters, pre-commercial thinners, tree planters, and other forestry workers had naturally become accustomed to responding to fires during the course of their work. A logger might accidentally cause a fire by his logging cable system or Mother Nature might decide to drop a few hot lightning strikes. These forestry workers were useful in the initial attack of small fires.

The Washington Department of Natural Resources and California Division of Forestry were blessed with large numbers of state convict crews to employ on the larger fires.

The Oregon Department of Forestry had a few prison convict crews. During up ticks in fire activity, the Department of Forestry began to contact forestry service contractors in the late 1980s to "borrow" forestry workers. Summertime was a slow time for many forestry service contractors as tree planting and prescribed burning were shut down. Over the next few years, NWFCA managed to get the Oregon Department of Forestry to set up an on-call fire crew contract for the state agency. There were about 10 to 15 fire contractors at the time. The Forest Service expressed a mild interest in the fire crew contract, but preferred to sit back and watch what might enfold. Many USFS and BLM government employees disliked contractors for a variety of valid reasons. Contractors themselves – who were in competition with each other - also disliked each other so there was much back-stabbing and lack of trust.

Contractors stumbled through the mid 1990s until the jackpot showed up in 2000. Surprised by the ferocity of the 2000 wildfire season which burned over 8 million acres, the Forest Service was forced to call upon private sector fire crews. Contractors made a financial killing in 2000 and the following years – such as 2002 – were just as beneficial. The looting of the federal treasury had begun. So, how does this relate to Porter's Five Forces?

Porter's model applied to wildfire contracting

Although one could certainly argue cogently that the success of contractors was not directly due to the use of Porter's Five Forces Model, there are a number of parallels (Figure 4-2). No offense intended to the National Guard crews or military units that were occasionally called up, but these crews tended to hang out on what are called "landings" and peer down off the side of the steep hill. Contractor crews and Forest Service crews were accustomed to running off the sides of the hills and were not afraid of getting their yellow fire shirts dirty. With recent events in Iraq, Afganistan, and who knows where the next US intervention, the "threat" of using US military or national guard units for fire fighting has simply disappeared. In terms of Porter's Five Forces Model, there has been a significant shift in the relationship of the Buyer (federal government) and the Supplier (contractors). Previously, the Buyer had lots of different suppliers in which to choose to dispatch to fires. However, the Buyer is now much more constrained and limited in choices.

Table 4.1: FY 2002 National Fire Plan Contracting by Program

Program	Fuels Treatment	Preparedness and facilities	Rehabilitation	Suppression	Total
BLM	$13.2	$3.1	$13.8	$6.5	$39.8
FWS	$0.4	$4.2	$1.6	$3.2	$9.6
NPS	$0.1	$0.9	$-	$1.6	$2.7
BIA	$14.2	$12.4	$2.2	$7.6	$36.5
DOI Total	$28.1	$20.8	$17.7	$19.1	$85.8
USFS	$10.5	$18.0	$13.1	$201.9	$243.6
DOI and USFS Total	$38.6	$38.9	$30.9	$221.0	$329.5

SUPPLIER POWER
Contractors worked together in an association to meet the common fire fighting standards and to create sufficient mass to attract the attention of the federal and state agencies.

BUYER POWER
Government needed large numbers of suppliers during peak critical periods and was unable to provide enough government crews.

RIVALRY

BARRIERS TO ENTRY
Contractors were required to show that they had crew bosses who had extensive experience and training.

DEGREE OF RIVALRY
The large concentration of Oregon contractors and fire workers improved the quality of the crews.

THREAT OF SUBSTITUTES
Government could use military and National Guard during peak fire seasons, but those crews performed poorly.

Figure 4.2: Five Forces Model related to fire fighting

The success of the early fire contractors lead to a mad stampede. In the year 2000, approximately 100 twenty person private fire crew were available for dispatch. As word spread about the looting of Uncle Sam's treasury, contractors rushed in after the almighty dollar.

Contractors switched from toiling on fuels treatment contract to wildfire suppression work because of the flood of federal dollars (Table 4-1). The Department of Interior and Forest Service spent a total of $38.6 million nationwide on fuels treatment through contracting, but over $221 million on fire suppression through contracting.

Contractors go where the money is. The scent of money was strong and palpable. In 2004, there are a reputed 365 fire crews as contractors chase after fire dollars. One of Michael Porter's Five Forces – Barriers to Entry – was not high enough to keep out contractors who were willing to cut corners and provide poorly trained crews. I'll be talking more about this problem in *Chapter 11: Limitations to job growth.*

In the next chapter, I'll be moving away from jobs and the economy to discuss our valuable natural resources in the Pacific Northwest. Our valuable forest natural resources offer us a unique opportunity to address serious environmental issues that are threatening their health while signficantly boosting the economy of the Pacific Nortwest.

Quickie

Our region needs to create competitive jobs based upon our unique resources. Porter's Five Forces Model provide a way to look at creating those types of industries and jobs that will be able to survive the rigors of global outsourcing.

Fire contractors are a home-grown example of how the Five Forces model can work successfully. Fire contractors became successful in Oregon due to a change in the ability of the Buyer (federal government) to obtain sufficient number of fire crews from both federal and non-federal sources.

Chapter 4

Our natural resources in the Pacific Northwest

Tracks of the white bear of enormous size...the men as well as ourselves are anxious to meet with some of these bear. The Indians give a very formidable account of the strength and ferocity of this anamal, which they never dare to attack but in parties of six eight or ten persons; and are even then frequently defeated with the loss of one or more of their party.

—Meriwether Lewis, journal entry, April 13, 1805

Flanked on one side by the Pacific Ocean, the Pacific Northwest region is split by the spine of the Cascade Range into two distinctive ecoregions. The Columbia River, in turn, creates the legal boundaries between much of Washington and Oregon. Washington is home to Bill Gates, founder of Microsoft, Inc. and Oregon is home to Phil Knight, founder of Nike, Inc. At last count, Bill Gates was far ahead of Phil Knight in net worth, which is probably reflective of the wealth division between the two states.

If the settlers had been wiser, the Oregon Territory should have been divided in half by the crest of the Cascade Mountains rather than the Columbia River (Figure 4-1). East-side residents of Washington have more in common with their fellow brethren of east-side Oregon than the urban populations of the Seattle-Tacoma megalopolis.

Geographic Features of Oregon and Washington

Figure 4.1: Geographic map of Oregon and Washington

Source: US Forest Service

In this chapter

This chapter is a primer on the forest natural resources of the Pacific Northwest. The chapter will describe the state of the forests, who owns the forests, and what is happening to the forests. This chapter will lay the groundwork for the discussions in the future chapters about how to create more jobs on public and private forestlands in the Pacific Northwest.

Everyone moves here for the quality of the environment

I was walking to the Eugene Public Library one day to "sneak" on their computers to work on this book (I am a resident of Springfield which is considered the bad side of the tracks in some parts of Eugene). As I passed one of the local organic food stores, a young bearded man with a clipboard asked me if I was a registered voter. I said yes. He then asked me if I would be interested in signing an initiative to ban clear cutting in the state of Oregon. I then asked him if he knew how many acres were clear cut on public lands last year. He paused for a second and said that he did not know. I told him that less than 500 acres were clear cut on federal forestlands last year in Oregon. I then asked him if he knew how many acres of public forestlands were in Oregon. He paused for a second and then said that he did not know. I told him that the BLM had approximately 2 ½ million acres of O & C forestlands and that the Forest Service had about 15 million acres of forestlands.

He then told me that the initiative did not deal just with public lands, but also with private lands. I then asked him if he knew how many acres of private forestlands were located in the state of Oregon. He paused for a second and then said that he did not know. I then told him that private companies owned 4 ½ million acres of forestlands and small woodlot owners owned 4 million acres. I then remarked that I did not think that the clear cutting of 500 acres on 17 1/2 million acres of public forestlands could be considered excessive.

I have no doubt that this individual was "well-meaning." In my previous life, I had also been "well-meaning" in my youthful days. The problem with "well-meaning" is that good works need to be tempered with adequate knowledge about our public and private forestlands.

The really big picture of the world's forests

For better or worse, since we're in the vaunted age of globalism, I'm going to drill down through a couple of layers to get to the forests of the Pacific Northwest. Here's the view from the Hubble Space Telescope.

Forests cover 30 percent of the land area worldwide, totaling about 10 billion acres (3.8 billion hectares). Forests in the United States account for only 6% of the world's forests. The bad news for forests worldwide is the destruction of 27 million to 49 million acres annually in the past decade, or approximately .3% to .6% annually.

Table 4-1: Population and forest area in 2000 and average annual change in forest area 1990- 2000

	Population		Forest Area	
Region	Total	Growth rate	Total	Annual change
	Millions	Percent	Millions hectares	Percent
Asia	3,634	1.4%	548	-0.1%
Africa	767	2.4%	650	-0.8%
Europe	729	0.0%	1,039	0.1%
North & Central America	478	1.6%	549	-0.1%
U.S. (by itself)	276	0.8%	226	-0.2%
South America	341	1.5%	886	-0.4%
Oceania	30	1.3%	197	-0.2%
World	5,978	1.3%	3,869	-0.2%

Source: 2000 USDA Forest Service RPA Assessment

Not surprisingly, human population has climbed 1.0% annually. Tropical forests in developing countries account for much of the decline of forests as the temperate forests in most developed countries are actually increasing in acreage (Table 4-1).

The global forest reserves are estimated to be 386 billion cubic meters, according to the United Nation's Forest and Agricultural Organization (FAO). The FAO estimates that 53% of the global use of wood is consumed as firewood while the remaining 47% is used in industrial production. In other words, poor people gather wood to cook their meals. Rich people convert trees into products.

The view down America's Main Streets

America's forests declined steadily by a total of 300 to 350 million acres during the major settlement period from 1800 to 1930. Between the 1930s to the 1960s, some 50 million acres of cleared farmlands in New England and worked-out cotton lands in the South reverted back to forestlands. Forestlands peaked at 761 million acres in 1963 and have declined slightly in the past thirty years (Table 4-2).

Americans like to consume wood products. Each American consumes about 1,800 lbs. of wood products per person annually. In 1996, more than 18 billion cubic feet of timber products were harvested in the United States with an estimated street value of $23 billion.

For the sake of argument, industrial timber products are the second most valuable agricultural "crop" harvested in the United States. Only corn has a higher annual crop value at $25 billion. However, timber used to be the highest value annual crop in the country, but dropped to second place after the drop in timber harvest in the Pacific Northwest.

Timber harvest in the Pacific Northwest accounted for 26% of the total volume harvested in the United States, but accounted for 40% of the stumpage value in 1986. Ten years later, Pacific Northwest timber harvest volume had declined to 15 percent of the United States total, but 24% fo the total stumpage value. In the Pacific Northwest area,

Table 4-2 Area of forestlands and timberland in the United States by ownership and region (millions of acres)

Ownership and region	1953	1963	1977	1987	1997
Forestland Ownership					
North	160	165	164	165	170
South	226	228	217	211	214
Rocky Mountain	141	140	138	139	143
Pacific Coast	227	227	224	221	219
Total	756	761	743	737	747
Timberland Ownership					
Public	145	146	138	132	146
Forest industry	59	61	68	70	66
Non industrial private	304	307	285	283	290
Total	508	515	492	486	503
Region -					
North	154	156	153	154	159
South	204	208	199	197	201
Rocky Mountain	66	66	60	61	71
Pacific Coast	83	82	79	73	72
Total	508	515	492	486	503

Source: 2000 USDA Forest Service RPA Assessment

the 1986 value of the timber harvest fell from 31% of the total value of all crops in the region to 21% in 1996 in the region. To understand forest acreage summarized in Table 4-2, note that while all timberlands are forestlands, not all forestlands are timberlands. Timberlands are forestlands considered to have an economic harvest value. Forestland and timberland ownership has stayed relatively consistent for the past fifty years. For example, total forestland ownership was 756 million acres in 1953, but had dipped only slightly to 747 million acres in 1997 (Table 4-2).

1.8 million people are employed in the primary and secondary wood products industry nationwide. The federal government defines 21 major manufacturing sectors within the United States under the North American Industrial Classification System (NAICS). Three wood products sectors are identified within the 21 major manufacturing sectors:

Table 4.3: Major wood manufacturing sectors

Sector	1996 value	Ranking
Paper Manufacturing	$153 billion	9th largest
Wood Products Manufacturing	$98 billion	12th largest
Furniture	$57 billion	18th largest

Source: 2000 USDA Forest Service RPA assessment

The quick tour of the world's and US' forests is finished. We can now focus on the forests of the Pacific Northwest.

Pacific Northwest Forests

This section will describe each of the major landowner types, timber harvests, management goals, and current conditions. It is important to understand the natural resources of the region so as to understand how to create jobs from the natural resources. Any discussion of the creation of jobs is difficult because of ecological differences, state differences, and ownership differences of the forests in the two states. Figures 4.3 and 4.4 may even confuse you somewhat more, but the two figures provides a general layout of the path that will be followed in this chapter.

Speaking with a broad brush, timber managers tend to use clear cutting on the west-side of the Cascades and tend

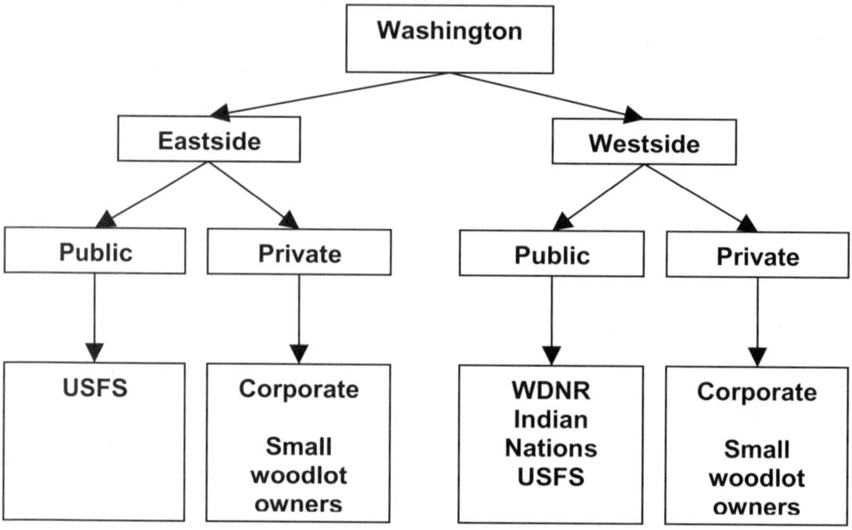

Figure 4.2 Primary landowners in Washington

to use partial cutting (removal of a selected number of trees) on the east-side of the Cascades. Private corporate timberlands in Oregon and Washington are generally maximized for timber production and are managed to generate the most yield. Non-industrial forestlands (commonly called small woodlot lands) may or may not be managed for timber production and may or may not be intensively managed.

State public lands managed by the Oregon Department of Forestry and Washington Department of Natural Resources tend to be intensively managed for timber harvest with consideration provided for other resources. While once being intensively managed, US Forest Service and BLM

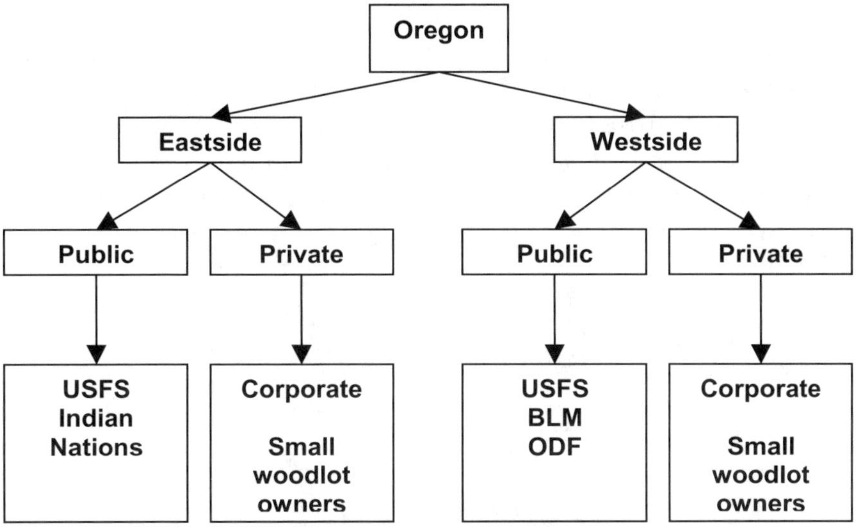

Figure 4-3: Primary landowners in Oregon

forestlands have been "hands off" for the past decade. Native American lands are managed both for timber resources and other values.

The different goals and management schemes of the myriad of owners provide for a much greater diversity of goods and services for both the environment and economy. For example, the public can not expect large private timber corporations to manage their lands for old growth characteristics and old growth species such as the spotted owl and red tree vole.

Public lands are set aside to protect those species which are more affected by timber harvests and other intensive management activities.

A look at Washington's forests

Of the 42.4 million acres of land in the state of Washington, approximately 21.8 million acres are classified as forestlands. Washington's forests are estimated to contain more softwood volume than any other state except for Oregon.

Washington forests are considered some of the most productive forests in the world. However, a number of studies have shown that there are substantial opportunities to increase intensive management activities that will generate more jobs and create more future timber supply. The composition of ownership of the forestlands in Washington is markedly different from Oregon. While the

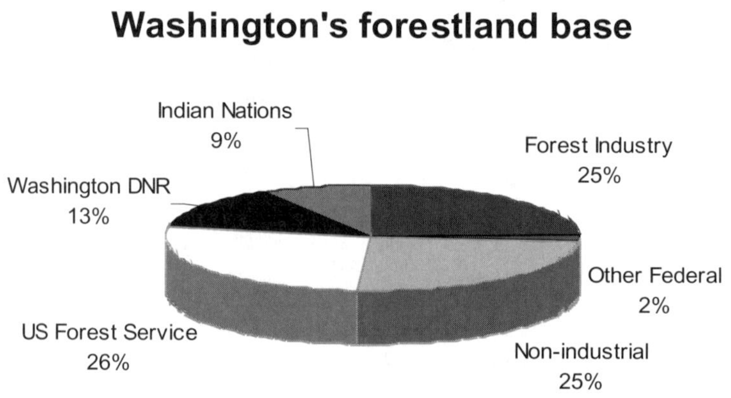

Figure 4.4 Washington's forestland base

Forest Service manages nearly 50% of the forest base in Oregon, the Forest Service manages only 26% of the forest base in Washington (Figures 4.4 and 4.8).

The private forest industry and small woodlot owners own nearly 50% of the forest base in Washington. In Oregon, the private forest industry and small woodlot owners own only 40% of the forestland base.

Tree types of Washington

Dominant and not so dominant tree types by acreage offer the reader an opportunity to understand the commercial and ecological values of the forests in Washington. Not surprisingly, Douglas-fir is the dominant species on over 8 million acres, far ahead of the next tree species - western hemlock - at 2 1/2 million acres (Figure 4.5). Of course,

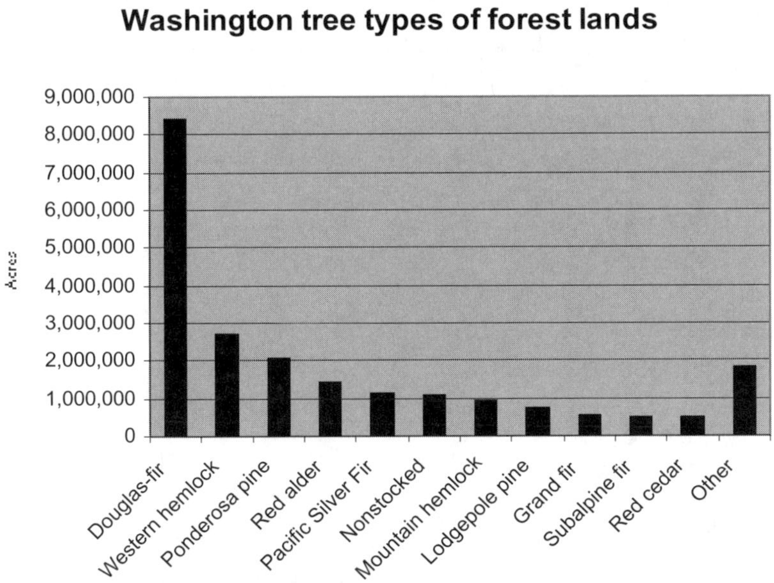

Figure 4.5 Major tree types of Washington's forests
Source: Washington Department of Natural Resources

one may argue that the tree types have been slightly altered by human activity. Most companies and government agencies tend to re-plant with Douglas-fir on the westside.

Washington's national forests

The Forest Service manages one quarter of the entire forestland base in Washington (Figure 4.4). The six national forestlands are located on the west side and northeast corner of the state (Figure 4-6). The Wenatchee National Forest is the largest and the Olympic is the smallest national forest in Washington. (Table 4-4). The Olympic National Forest is one of the rainiest national forests in the entire nation and surrounds Olympic National Park. The Mt St Helens National Monument is a separately designated unit located within the Gifford Pinchot National Forest.

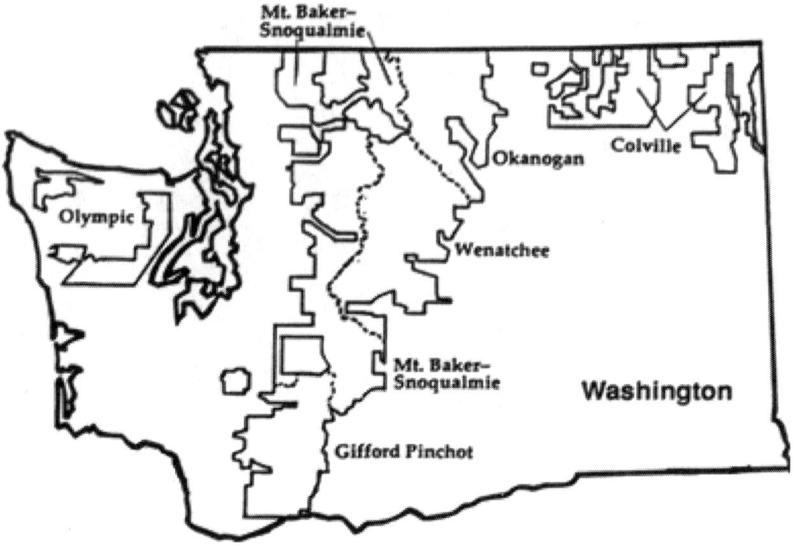

Figure 4.6 National forests in Washington
Source: USFS

Table 4-4: Washington national forests by acreage	
Colville	1,080,000
Gifford Pinchot	1,050,000
Mt Baker Snoqualmie	980,000
Okanogan	945,000
Olympic	535,000
Wenatchee	1,190,000
Source: USFS	

Washington Department of Natural Resources

Unlike the Oregon Department of Forestry which owns and manages less than 800,000 acres, the Washington Department of Natural Resources (WDNR) manages 2.1 million acres of forestlands. Washington counties and municipalities manage less than 200,000 acres of forestlands.
The Washington DNR also manages 2.4 million acres of tidelands, shorelands, and beds of navigable waters as

well as 800,000 acres of grazing lands. The agency is responsible for protecting 12 million acres of private and state owned forested acres from wildfires.

WDNR is responsible for administering forest practice rules which guide harvesting and road-building on private forestlands. WDNR generates approximately $300 million annually primarily through the sale of timber. The funds are dedicated to universities and schools. WDNR timber harvest has averaged approximately 600 million board feet for the past five years.

Washington's private industry lands

Washington's forest industry is the second largest in the country - behind the state of Oregon. Washington's forest industry produces 12% of the total US softwood lumber production and approximately 5% of the total US plywood production. Harvest levels on private industry lands (both

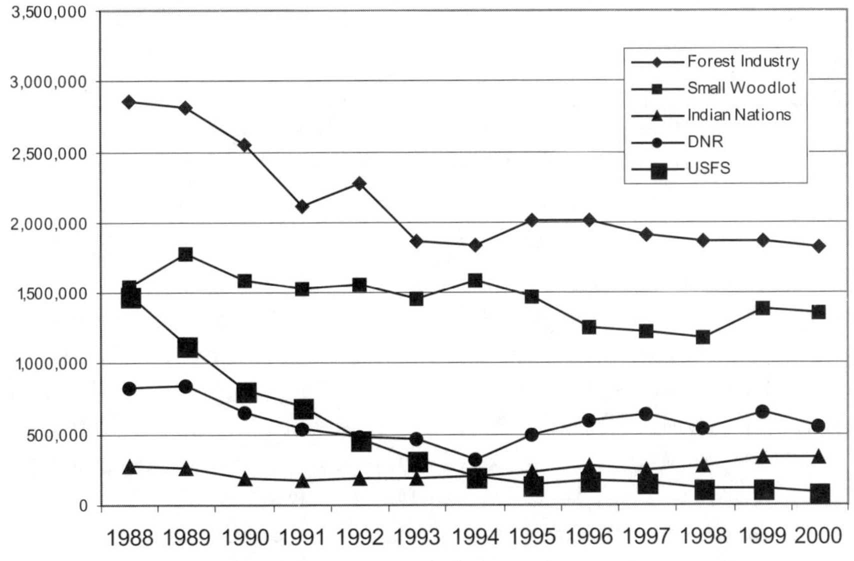

Figure 4-7: All ownerships timber harvest 1988 - 2000
Source: Washington Department of Natural Resources

industrial and non-industrial) have ranged from 3.0 - 3.2 billion board feet over the last eight years (Table 4-5).

Table 4-5: Timber Harvest: All Ownerships 1988-2000
(thousands of board feet)

	Forest Industry	Small Woodlot	Indian Nations	DNR	USFS	Total
1988	2,864	1,541	270	826	1,485	7,045
1989	2,807	1,774	261	841	1,141	6,850
1990	2,550	1,596	182	657	817	5,849
1991	2,117	1,532	171	535	703	5,103
1992	2,276	1,566	185	476	461	5,017
1993	1,861	1,460	192	461	322	4,329
1994	1,834	1,586	198	322	199	4,155
1995	2,015	1,474	230	496	150	4,392
1996	2,017	1,255	269	599	181	4,366
1997	1,917	1,221	252	645	166	4,246
1998	1,867	1,176	275	546	110	4,021
1999	1,864	1,381	333	662	116	4,382
2000	1,824	1,352	330	559	81	4,176

Source: Washington Department of Natural Resources
An "other" category was deleted because of the small amounts.

For those such as myself that can't see the forest trend for the sheer confusion of numbers in Table 4-5, Figure 4-7 graphically displays the timber harvest trends. I deleted the other ownership category because of its miniscule amount. The US Forest Service has dropped from a timber harvest of nearly 1.5 billion board feet in 1988 to less than 100 million board feet in 2000. Even the forest industry shows a considerable drop from 2.8 billion board feet in 1988 to 1.8 billion board feet in 2000. That translates into lots of lost jobs – notwithstanding the substantial productivity gains. The decline on private forest industry land likely reflects a drop in the available harvest stock.

A look at Oregon's forests

Oregon is a big state. Of the 61.4 million acres of land in the state of Oregon, 27.5 million acres are classified as forestlands. Oregon's forests are estimated to contain more total softwood volume than any other state in the union. Approximately 24.5 million acres of Oregon's forestlands are capable of producing 20 cubic feet or more per acre per year, which used to be the common definition for commercial forestlands. Oregon's forestlands are both publicly-owned and privately-owned. The ownership patterns data from the late 19th and 20th century. Lands were given by the federal government in the 19th century to pioneers. In the 20th century, the government set aside lands for national parks, forests, and rangelands.

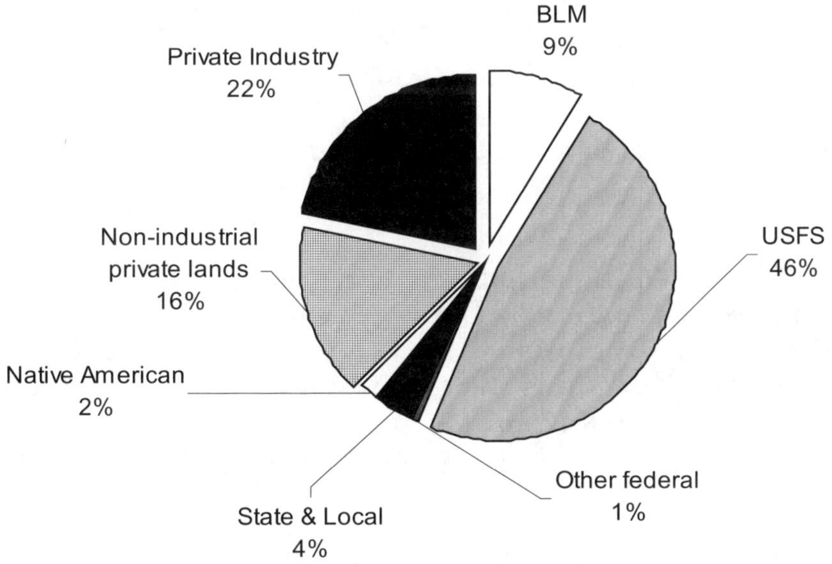

Figure 4.8 Oregon's forestland ownership base

Big Kid on the block - Forest Service

The Forest Service manages nearly 50% of the entire forestland base in Oregon (Figure 4.8). National forests are generally found at higher elevations than other ownerships.

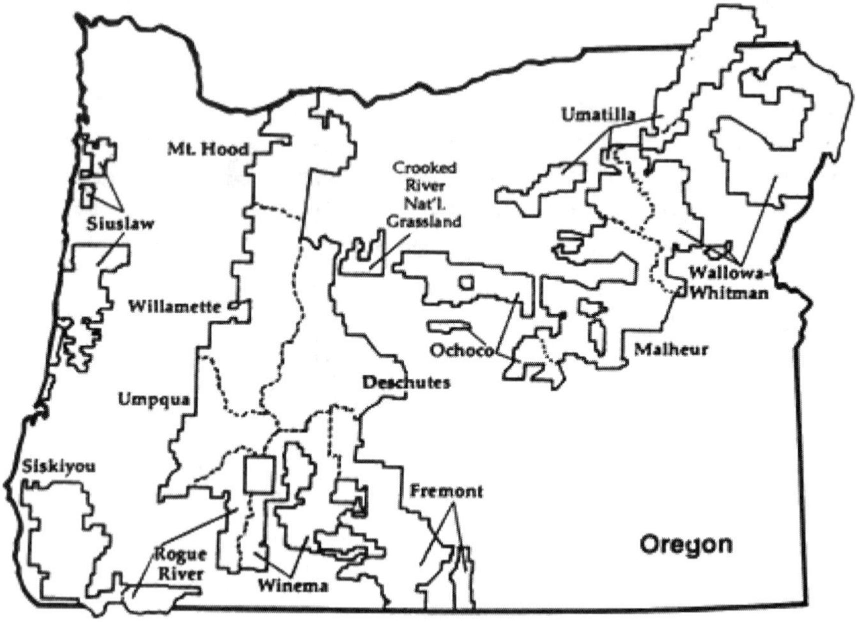

Figure 4.9 Oregon's national forests
Source: US Forest Service

The Wallowa Whitman National Forest in northeast Oregon is the largest national forest in Oregon with over 1.4 million acres. By contrast, the Rogue River National Forest in southwest Oregon is the smallest national forest at less than 550,000 acres (Figure 4.10). Most Oregonians are located within a one hour to one and half hour drive to a national forest (Figure 4.9).

The US Forest Service maintains plots throughout Region 6 (Oregon and Washington) to measure the amount of tree growth. Watching trees grow is about as exciting as watching paint dry. In our busy and frenetic days, we don't really grasp that the fifteen-year old Douglas-fir tree in our backyard has grown two feet in the past year. Forest Service plots suggest that the Region 6 forests grow approximately 12.4 billion board feet of growth each year. How much is 12.4 billion board feet of growth? A board foot is 12" long by 1" deep by 1" wide. There are approximately 5,000 board feet that can be carried on a log truck. If all of the growth were to be removed from the Oregon and Washington national forests each year, one would need to hire 2,480,000 log truck trips each year to remove the growth.

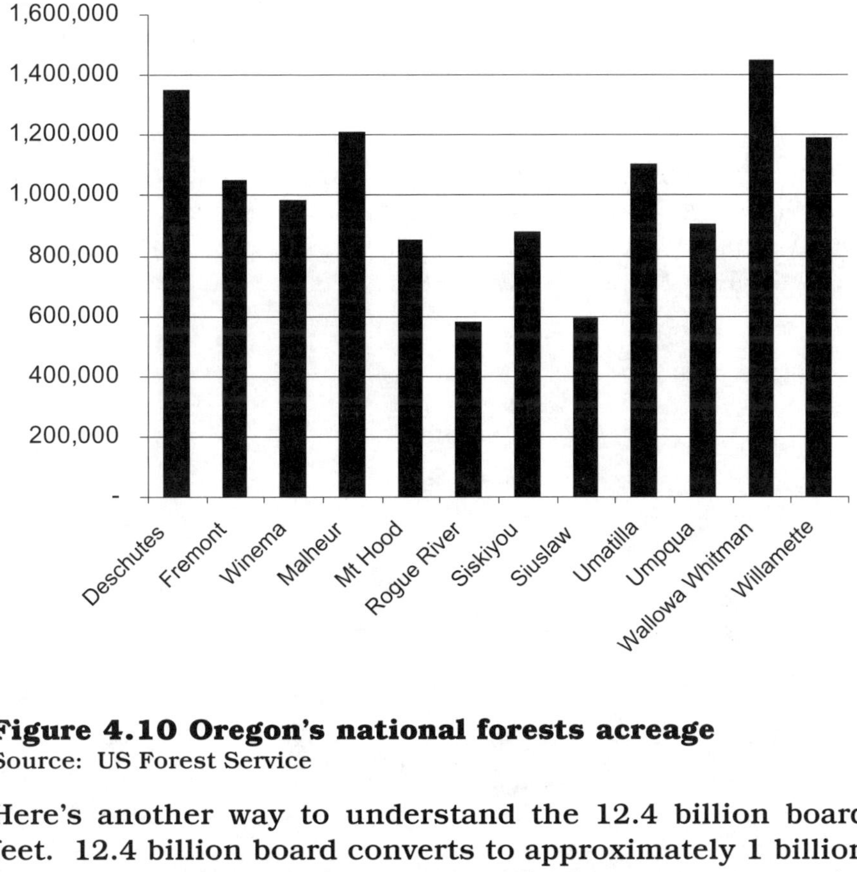

Figure 4.10 Oregon's national forests acreage
Source: US Forest Service

Here's another way to understand the 12.4 billion board feet. 12.4 billion board converts to approximately 1 billion cubic feet. One acre is equal to 45,360 square feet. An acre is approximately the size of a football field. As James Peterson of *Evergreen Forestry* describes, visualize a football field full of wood stretching 4 miles into the air. That is what is grown each year on national forests alone in the Pacific Northwest.

Trees are also dying each year. The tree mortality is projected to be 1.75 billion board feet each year on our national forests alone in Oregon and Washington. Using our standard measurement of 5,000 board feet on a log truck, 350,000 log truck trips would be needed each year to remove what had died in the forest that year in the Pacific Northwest.

Back to our football field metaphor. The mortality of 1.75 billion feet converts into one mile high of a football field size mass stretching into the sky. That is just what is dying each year of wood.

However, mortality varies quite significantly in different parts of the Pacific Northwest - particularly between east side and west side. Mortality for three national forests in northeast Oregon is a much higher percentage of the growth than for west-side forests (Figure 4-11).

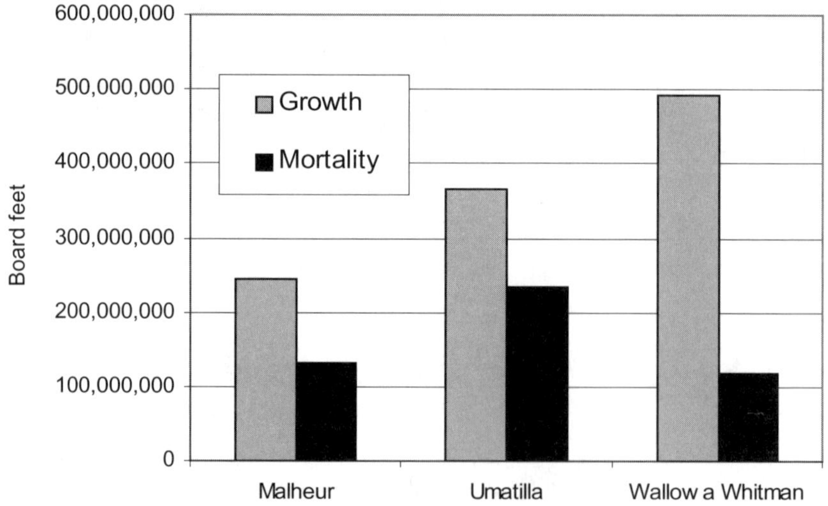

Figure 4-11 Sample growth and mortality
Source: Unpublished data - USFS Region 6

Like the financial ads say about buying stocks, the actual growth and mortality will vary from year to year based upon rainfall, insects, disease, fire, and other environmental factors.

The Big Kid's little brother - BLM

The BLM manages more land than the Forest Service nationally. The BLM is responsible for management of 262 million acres – nearly one-eighth of the land mass of the United States. Most of the lands are grasslands, tundra, desert lands, and other lands that generally have low economic value - except for 2.6 million acres of forestlands in western Oregon.

The O & C lands for timberlands that were revested from an Oregon and California private railroad that failed to comply with its contract. These timberlands generally have very high site productivity in western Oregon. These lands are found at the lower elevations of the Cascade and Siskiyou Mountains and are usually intermixed with private industrial corporate lands.

The O & C lands were the golden goose for counties in the 1970s and 1980s. The federal government handed over 50% of the timber sale receipts from the sale of timber in those counties.

Other federal agencies with forests include National Park Service (174,000 acres) and US Fish and Wildlife Service (4,000 acres).

Indian Nations lands

The Indian Nations, of which the largest is the Warm Springs Indian Reservation, manage 416,000 acres of forests in the state. The Warm Springs Indian Reservation lands are located northwest of Bend along the east flank of the Cascade Mountains. Indian Nation lands used to be classified as federal government lands, but many timber books are now identifying the lands as private industrial.

Oregon Department of Forestry

State agencies and other counties and municipalities manage less than 4% of Oregon's forests. The largest state forest land management agency is the Oregon Department of Forestry which is responsible for 596,000 acres plus 183,000 acres for the State Land Board. Most of the Department of Forestry lands are located in the Tillamook State Forest in northwest Oregon and the Elliot State Forest near Coos Bay.

Like the Washington Department of Natural Resources, the Oregon Department of Forestry is responsible for enforcement (or lack thereof) of the state forest practices act. Timber companies and small woodlot owners are required to seek permits for logging, herbicide application, and other types of intensive management activities.

West-side private industry lands

Private forest industry and non-industrial forest landowners account for over 36% of the forestlands, totaling 9.3 million acres, in Oregon. The site productivity in western Oregon is quite high on private forest industry lands.

For this section about trends in the Oregon private forestry sector, I am relying on a number of studies by Oregon State University forestry professor Darius Adams. Professor Adams is a major contributor to the annual national Resource Planning Act timber assessment by the federal government. Professor Adams, Randall R. Shillinger, Greg Latta, and Adrienne Van Nalts published a study entitled: *Timber Harvest Projections for Private Land in Western Oregon* in May 2002. The study assessed the future harvest potential and future conditions of the private forestry sector base in western Oregon.

The study noted that 80% of the timber harvest in Oregon is produced from private lands. The study identified 4.34 million acres of private industry with a growing stock inventory of 12.1 billion cubic feet. Non-industrial private forest lands amounted to 1.93 million acres with 4.65 billion cubic feet of growing stock inventory. Like the federal deficit, numbers can be both confusing or intimidating. The point to remember is that Oregon contains more softwood tree volume than any other state in the country. So, just remember that those are big numbers.

The growing stock has declined from approximately 17.5 million cubic feet in 1950's to 12.1 billion cubic feet in the 1990s in western Oregon. The private industry harvested virtually all of its old growth and replaced those forests with much faster growing second growth stands. 90% of the trees on its lands are less than 60 years of age.

Non-industrial private land (NIPF) stock inventory has declined from approximately 7 billion cubic feet in 1950 to 4.6 billion in the late 1990s in western Oregon. Somewhat similar to private timber lands, 90% of the NIPF tree inventory is less than 80 years of age.

The Adams et al study projected that forest industry lands could maintain the recent levels of harvest for the next fifty years. The declining trend of the last four years would be abated.

Table 4-6: Oregon's 25-year harvest

Year	Industry	Small wood-lot	Native American	State	BLM	USFS	Total
1977	3137	453	115	228	1021	2913	7867
1978	3163	387	121	235	837	3232	7975
1979	2953	255	111	223	956	3167	7665
1980	2930	204	105	186	797	2399	6621
1981	2522	180	95	216	677	1981	5671
1982	3221	219	126	175	312	1688	5741
1983	3124	249	112	257	789	2902	7433
1984	2768	310	101	249	920	3164	7512
1985	3009	323	121	268	891	3480	8092
1986	3066	428	104	225	1042	3850	8715
1987	2791	490	117	199	1115	3451	8163
1988	2673	586	121	270	1439	3487	8576
1989	3027	694	124	198	1026	3307	8376
1990	2628	601	98	137	704	2014	6182
1991	2817	494	87	91	486	2068	6043
1992	2751	830	111	135	483	1403	5713
1993	2700	909	75	116	361	1102	5263
1994	2471	773	80	130	92	596	4142
1995	2736	696	79	109	139	515	4274
1996	2463	555	71	114	289	401	3893
1997	2653	480	79	176	136	523	4047
1998	2470	370	71	141	122	333	3507
1999	2555	459	68	246	150	233	3711
2000	2712	455	62	255	83	245	3812
2001	2624	281	63	268	38	135	3409
2002	2985	333	71	269	55	167	3880

The "other public" column has been deleted because of its small amounts

Source: Oregon Department of Forestry

The study indicated lots of uncertainty over non-industrial private forest lands in western Oregon. Due to the large amount of trees growing on non-industrial private forest lands, the study indicated that the annual timber harvest levels could double and maintain even flow.

East side private industry lands

The picture of private forestlands on the east side of the Cascades is not so pleasant. Reflecting the same problems facing the national forest lands in eastern Oregon, the private forestlands have been suffering from insects and disease.

Similar to public lands, the private lands in eastern Oregon have been hit hard by spruce budworm, Douglas-fir tussock moth, and other bark beetles and defoliators.

The private industry owns approximately 1.6 million acres with a standing volume of 1.786 billion cubic feet. The projected harvest is expected to drop a steep 50% in the coming years because the industry has been heavily harvesting its own lands due to mortality losses from diseases and insects.

The non-industrial private forestlands total 1.1 million acres with a standing volume of 1.655 billion cubic feet. An unpublished draft report entitled *Private Timber Harvest Potential in Eastern Oregon*, dated May 2002, by Professor Darius Adams and Gregory Latta projected that even flow harvests from the lands could be increased by 20% from the historical average.

Lumber production from all ownerships (including federal lands) in eastern Oregon declined from 1.8 billion board feet in the 1980's to .9 billion board feet in 1999. The number of mills collapsed from 42 in 1988 to 14 in 1998.

Small woodlot owners in Oregon

Non-industrial forest landowners, commonly called small woodlot owners, own approximately 4.4 million acres of forestlands in Oregon. Of this total, approximately 2.6 million acres are considered commercial forestlands. Oregon has approximately 25,000 small woodlot owners. The median age is 55 years of which 25% are older than 65 years. About 1/3 of the owners are employed in wage and salary jobs, 1/3 are self-employed, and 1/3 are retired. The pre-tax household earnings for 33% of the landowners is $10,000 to $30,000 based upon 1989 data. An additional 33% of small woodlot owners earn $30,000 to $50,000. Approximately one half of the landowners live on their lands or have a second home on the land.

According to an unpublished survey of landowners conducted by the Department of Forestry, income from timber is important to 50% of the small woodlot owners. Larger landowners are more likely to harvest and are more likely to invest in forest improvements for forest productivity. The study found that most of the non-industrial forestlands is owned by a small number of landowners. Landowners who have more than 160 acres (17 percent) account for 71% of the non-industrial forestland base.

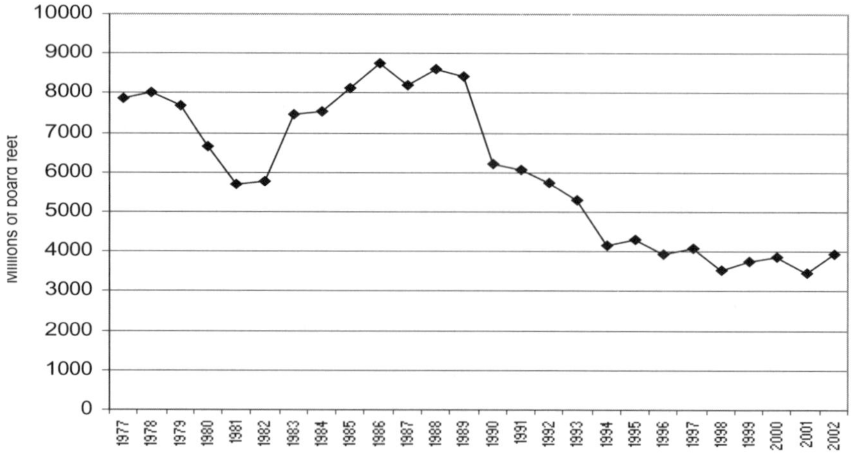

Figure 4-12 25-year Oregon timber all ownerships harvest
Source: Oregon Department of Forestry

Oregon's overall harvest activities

The harvest levels typically ran 7 to 8 billion board feet in the 1970's and 1980s (Table 4-6). The timber level dropped to 6.04 billion board in 1991 and has trended downward below 4 billion in the last several years (Figure 4-12). Both Oregon and Washington harvest levels (Figures 4-7 and Table 4-6) show big plunges in federal timber harvest.

I feel compelled to include Table 4-7 for my friend carrying the petition to ban clearcutting in Oregon. What the table clearly shows is that clearcutting has literally disappeared as a harvesting technique from federal public lands. Only 500 acres were clearcut on 17 1/2 million acres of federal forestlands.

Table 4-7: Harvested acres in Oregon - 2001

Ownership	Clearcut		Partial cut		
	West	East	West	East	Totals
BLM	115		2,841	512	3,468
Native American	638	184	152	3,038	4,012
Industry (1)	88,912	13,082			101,994
Local Government (1)	793	0			793
Non industrial private (1)	17,379	589			17,968
State	2,533		14,113	514	17,160
US Forest Service	158	500	3,788	29,274	33,720

1) 2001 data is not available on partial cut acres for Industry, Local Government or non-industrial private ownership groups. (Source: Oregon Dept. of Forestry).

What about our big brother to the south?

California has suffered the same fate as Oregon and Washington although one could argue that the situation is more serious in California. The California Board of Equalization tax records show that only 1.66 billion board feet of timber was harvested in 2003, down from over 4.67 billion in 1988. Logging dropped 40% and 90% on private and public lands respectively. In 1988, the state produced nearly 75% of the timber gross that was consumed in the state. By 2003, the figures were nearly reversed as California now imports 70% to 80% of its wood from other states and overseas. Of the 130 sawmills existing in 1988, 50 of those sawmills have closed. The closest sawmill to the massive stand of dead and dying trees on the San Bernardino Mountains near Los Angeles is 250 miles away. The California Forest Products Commission estimates that 15,000 forest jobs have been eliminated since 1998.

Quickie

Oregon and Washington are No. 1 and No. 2 respectively in the total amount of softwood volume of any state in the Union. The Forest Service manages nearly 50% and 25% respectively of the forests in Oregon and Washington. Forest Service and BLM harvests have dropped approximately 90% from 1988 in Oregon and Washington. Private industry harvests have declined in Washington in the past fifteen years, but have held relatively steady in Oregon.

Chapter 5

What's Happened to the Forest Service and BLM?

"The increase in forest density is astonishing. In some areas, ponderosa groves that once held no more than a dozen or two large trees now contain more than 2,000 stems per acre. Picture a solid block of wood the length and width of a football field stretching a mile into the sky. That's how much new wood fiber is added to Arizona and New Mexico forests every year. The block in your mind's eye contains 185 million cubic feet of wood, enough to build 1,800 square-foot homes for 90,000 families."

Jim Petersen, writer and executive director the non-profit Evergreen Foundation, from "The New Pioneers", Evergreen Magazine

A long time ago in a universe far and far away, I was working at the Condon Guard Station on the Flathead National Forest in western Montana as a seasonal firefighter. The Bob Marshall Wilderness lay immediately to our east and the Mission Mountains Wilderness lay to our west. The Condon Guard Station was a great place to watch the alpine glow on the peaks as the sun set.

Our supervisor had assigned us the task of painting all of the eight buildings at the guard station. We fired up the paint spray machine during our first week and began to spray paint the first two-story dormitory in brown paint.

The supervisor came out during the middle of the first day to see that we had already painted one whole side of the building. We could tell that he was quite disturbed. He took away our paint spray machine and gave us small paintbrushes. Although he never gave us a reason, we assumed that he had calculated that we would have finished all of the buildings within a couple of weeks and would not have any more work to do for the rest of the summer. Efficiency was not necessarily a priority for Forest Service fire crews. In hindsight, those were the glory days of the Forest Service when the agency had hundreds of millions of dollars to spend on firefighters like me to paint buildings.

In this chapter

Let's step a few years back to discuss what has happened to the public natural resource agencies during the tumultuous period of the last twenty years. The reason why we need to look back is to understand that the Forest Service and BLM used to generate an enormous number of jobs, both as direct hires working for the government as well as contractors working on its lands. Both agencies had quite a positive financial impact on our Pacific Northwest region, in particular, the rural areas of our states.

Big battles in the 1970s and 1980s

Back in the days of the 1970s and and 1980s, big forest battles roiled through the Pacific Northwest as the timber industry and environmentalists squared off against each other. At stake was a very important prize to both parties. Four billion board feet of timber flowed

off federal lands in Oregon and Washington each year representing 15% of the total timber consumed in the United States. The forest products industry was intent on protecting its financial investments over the past seventy years. For environmentalists, the increasing conversion of old growth forests to managed forest plantations represented a dagger in their heart each time that a new timber sale was sold and harvested.

The Forest Service used to brag that it was the only federal agency in the entire federal government that was a net contributor to the federal treasury. The Forest Service made money through the sale of timber.

Dick Worthington, USFS Regional Forester in the 1980s, was well known politically throughout the region. I used to compare the power of the regional forester to that of the respective governors of Oregon and Washington. In some senses, the Regional Forester carried more power as he managed nearly 25 million acres in Oregon and Washington. Forest Supervisors were equally well known in their local areas. Articles about the Forest Service and BLM would appear several times a week in the daily newspapers. What happened on federal lands had a big impact on the economy of the two states and the number of timber sales sold annually on public lands had a big impact on local county budgets.

How the Forest Service operated in the pre-owl days

It's important to understand how the Forest Service operated in the "pre-spotted owl" days of the 1970s and 1980s. Timber sale dollars didn't flow directly out of the region back to the evil people in Washington, DC. The Forest Service turned over 25% of its timber sale receipts to the local counties while the BLM provided a generous 50% of its timber sale receipts to the local counties. As one federal forester explained to me, Uncle Sam was "Uncle Sugar." The Forest Service set aside money from the timber sale receipts for what was known as Knudsten-Vandeburg (K-V) funds to properly reforest the lands, and for brush disposal funds (BD funds) to treat fuels in the timber sale boundary. During the spring, summer and fall months, the Forest Service hired large numbers of seasonal forestry workers to manage and protect the lands. In its heyday, most of the national forests in Oregon and Washington operated in the black. Each of the national forests could plow money back into the local economies either through the timber tax receipts, hiring local residents to work on the national forests, or hiring contractors from its huge K-V funds.

Big battles were fought over wilderness bills in the early 1980s and the forest plans in the mid-1980s. I was one of those folks who worked for the Oregon Wilderness Coalition for passage of a one million acre wilderness bill in Oregon because there was a need to slow down the monster. The problem was that the timber sale program needed to be reduced by a factor of 15% to 25%, not eliminated.

The 1980s and 1990s have not been kind to the Forest Service or the BLM. The agencies faced an increasingly hostile and polarized public about the level of the timber sale. Internally, the Forest Service and BLM were attempting to add more females and minorities to its leadership ranks. Well-qualified white males were routinely passed over for less-experienced females and minorities in the 1980s and the early 1990s. Environmentalists caustically referred to the BLM as the "Bureau of Livestock and Mining" while the timber industry caustically referred to the BLM as the "Bureau of Lazy Men." Times were tough for both agencies.

Mental and physical breakdown of both agencies

The mental and physical collapse of both agencies seems to have happened in the period of time from 1990 to 1993.

In the space of three years, the combined timber sale program of both agencies dropped from 4 billion board feet to less than 1 billion board feet. Federal judges with little to no natural resource understanding began to routinely enjoin timber sales under a belief that these judges were courageously protecting the environment. Experienced foresters began to drop out as the Forest Service began a formal reduction in force. What used to be a fairly well-oiled operation broke down under the weight of conflicting opinions as to the mission of both agencies. In recent years, the combined timber harvest level has dropped below 400 million board feet in Oregon and Washington.

Figure 5-1 doesn't really give justice to the effects of the crash. There were real human beings who were working

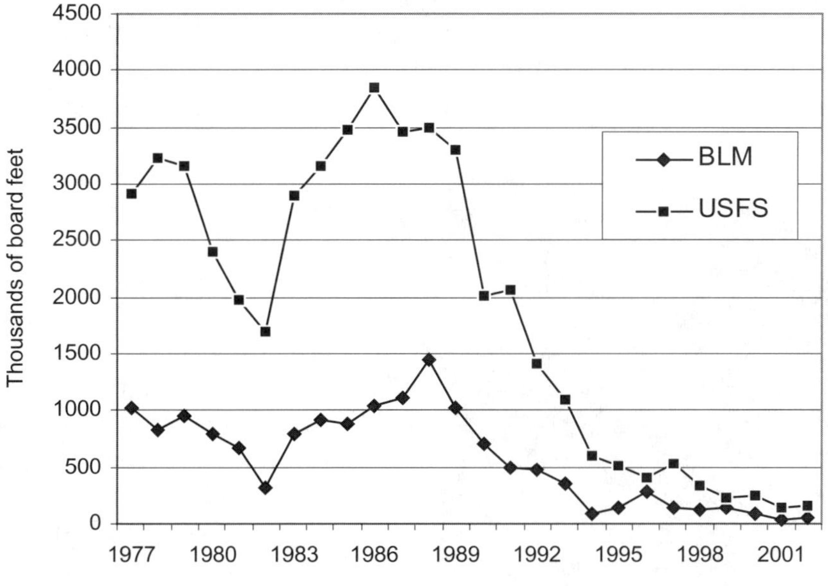

Figure 5.1 Collapse of the Forest Service and BLM timber harvest in Oregon
Source: Oregon Department of Forestry

– both in the public and private sector – who are no longer working in the woods.

With the collapse of the federal timber sale program, sawmills that were dependent upon federal timber began to close. In the last thirty years, approximately 800 sawmills have closed in the western United States with a significant number of those mills located in Oregon and Washington. While there was a natural consolidation that was occurring in the industry due to automation that would have closed many of the sawmills, a number of the mills simply closed for lack of a stable and predictable timber supply in the region.

Fast forward another ten years

Although it was only early July in 2003, the Clark Fire on the Willamette National Forest rapidly grew from Fall Creek and roared through old growth forests. Both the Forest Service and Department of Forestry dispatched crews to battle what would eventually become a 5,000 acre fire. The fire was stubbornly difficult to suppress because of the large amounts of biomass and the steepness of the terrain.

During the two weeks that the fire was belching smoke, our local newspaper – *The Register Guard* - ran a front page story entitled *Clark Fire – Let it Burn.* The news reporter quoted a former Forest Service firefighter. I recognized his name because he had written a "let burn" article for my University of Oregon alma mater magazine, *Old Oregon.* Although the news reporter was careful to simply quote the comments of the former wildland firefighter, it was quite instructive that the *Register-Guard* would even print such silliness in its paper. I talked with one Department of Forestry fire boss about the article and joked that letting the Clark Fire burn naturally would likely pump smoke into the Willamette Valley for the next three months as the fire chewed through 100,000 – 200,000 acres of old growth and plantation forests.

As one Canadian friend rhetorically asked me about Americans, "Where's the common sense?" Of course, there are two reasons why not to place too much credence in his words. First, he's a Canadian. Second, he's copublisher of *Logging and Sawmilling Journal.* But I had better not dismiss the Canadians so quickly. They've managed to convince us to import 1/3 of lumber consumed in the United States and hold a $57 billion trade balance with our country.

Quickie

The Forest Service and BLM once provided tens of thousands of jobs in the Pacific Northwest on their lands both through the public sector and the private sector up until the early 1990s. Timber sale receipts helped fund schools, roads, and local county governments. Those jobs have mostly disappeared and the timber sale receipts have been replaced with substituted tax dollars that are coming from our increasingly bankrupt federal government. To rebuild the economy in the Pacific Northwest, we need to rebuild the Forest Service and BLM.

Postscript

Our family visited Glacier National Park in western Montana two years ago. On our journey back, we stopped off at what used to be my old Condon Guard Station. The station was no longer occupied by the Forest Service. Rather, the station was being used by a local ecological community group who had asked to use the station after it was "abandoned" by the Forest Service.

Chapter 6

The decline in health of our public and private lands

At the time of creation, the Creator placed us in this land and gave us the voice of this law and that is our law. Ultimate sovereignty is vested in the people, who received their sovereign authority in the form of laws given by the Creator and by the land itself. We shall, as we always have, live in balance with the land and never use more of our precious natural resources than can be sustained forever.

Tribal perspective on Resource Management
From the Declaration of Sovereignty
Confederated Tribes of Warm Springs

I was visiting Dan Robertson, president of our Northwest Forestry Contractors Association, at his Coos Bay office about the outlines of this book. While discussing the conditions of our public lands, Dan noted that he had worked for three summers as a forester trainee on the Baker Ranger District of the Wallowa Whitman National Forest while attending Oregon State University School of Forestry. The direction at the time was to maintain the large, open old growth ponderosa pine characteristics of the district. He remarked that when he last visited the Baker Ranger District, the forests were choked up with many competing and stunted trees. He was disturbed with the condition of the forest.

In this chapter

I know that I bored you with my lecture about Porter's Five Forces in Chapter 3 about how to create jobs. Unfortunately, I'm going to have to bore some of you for a primer on fire regimes and condition classes. If you know the five fire regimes and three current condition classes, jump ahead to the section entitled **Condition of national lands**. If you're still reading, I'll try to make this as painless as possible. You might be thinking that I am drifting away from the *How to create good jobs in the Pacific Northwest* purpose of this book, but hopefully I'll be able to tie ecology and economy back together again in the later chapters.

Ecosystems are fire-adapted

Most ecosystems in this country are fire-adapted. Many species are dependent upon fires. For some ecosystems such as the chaparral in southern California, fire made a frequent and regular appearance. Fires, whether started by the native Americans or by lightning, frequently affected the lands of eastern Washington and Oregon. For example, certain types of softwood cones such as lodgepole pine need the heat from fire to open.

Natural fire regimes

As you are aware, the federal government and universities spend princely amounts of money studying everything imaginable under the stars. Although not equal to research on defense spending, the government has spent a few dollars to classify and map ecosystems and their fire regimes. One of the most significant efforts was published by the USDA Rocky Mountain Research Station in a report entitled *Development of Coarse-Scale Data for Wildland Fire and Fuel Management* by Kirsten Schmidt, James Menakis, Colin Hardy, Wendell Hann, and David Bunnell. The report is located at www.fs.fed.us/rm/pubs/rmrs_gtr87.html. While the title is a bit confusing, the maps and associated data are eye-popping. The report displays the conditions of all of the lands in the United States in regards to their departure from the natural ecosystem. The purpose of the report is to display whether the ecosystems are at fire risk for major change. The authors developed five historical fire regime classifications:

Table 6-1: Historical natural fire regimes	
Code	Description
I	0 – 35 year frequency (a), low severity (b)
II	0 – 35 year frequency, stand-replacement severity
III	35 – 100+ year frequency, mixed severity
IV	35 - 100+ year frequency, stand-replacement severity
V	200+ frequency, stand-replacement severity
a Fire frequency is the average number of years between fires.	
b Severity is the fire effect on the dominant overstory vegetation.	

What is important to understand about these five codes is that the researchers were trying to develop "coarse" modeling of data across the landscape of the United States to compare the degree of change from pre-white settlement periods.

A classic example of a natural fire regime with a Code I is an east-side, ponderosa pine forest. Lightning fires typically burned beneath the old growth ponderosa pines every 7 to 25 years. Frequent fires prevented shade-tolerant firs from getting established in this zone. This zone favored large, more fire-resistant ponderosa pine trees.

A classic example of a Code V, 200+ year fire frequency is an old-growth, coastal Douglas-fir or Sitka spruce site that did not experience fires except when conditions became extremely dry which occured at quite lengthy intervals.

Finally, I would like to make one confusing note about the severity effect of fire on the dominant overstory vegetation. While we normally think of "overstory" as trees, overstory can also be brush as in the example of the chaparral hills of southern California if that is the dominant species. If a fire burns through the "brush overstory," the fire would be classified as a "stand replacement."

The key to remember is that the authors have entitled this report "coarse scale." The authors are attempting to reduce lots of environmental variability into five fire regime types. These natural fire regime categories are some simple, basic generalizations about the frequency of fire in a particular ecosystem.

So, here's where it will start to get a little tricky when we move to what are called the condition classes. Condition classes are inter-woven into natural fire regimes.

Condition classes affect fire regimes

The researchers who drafted the *Development of Coarse-Scale Data for Wildland Fire and Fuel Management* study then attempted to describe the degree of departure from the historical fire regime. There are many different reasons why the present condition has departed from the historical regime such as fire suppression, timber cutting, grazing, introduced plant species, introduced insects and diseases. For example, fire suppression has been blamed for allowing lots of fuels to accumulate on the forest floor and in the canopy. The authors of the study created three condition classes with the following definitions and descriptions (which I have included as written by the Forest Service authors).

Condition Class I:
Fire regimes are within an historical range, and the risk of losing key ecosystem components is low. Vegetation attributes (species composition and structure) are intact and functioning within the historical range.

Example of a management option for Condition Class I: *Where appropriate, these areas can be maintained with the historical fire regime by treatments such as prescribed fire use.*

Condition Class II:
Fire regimes have been moderately altered from their historical range. The risk of losing key ecosystem components is moderate. Fire frequencies have departed from historical frequencies by one or more return intervals (either increased or decreased). This results in moderate changes to one or more of the following: fire size, intensity and severity, and landscape patters. Vegetation attributes have been moderately altered from their historical range.

Example of a management option for Condition Class II:
Where appropriate, these areas may need moderate levels of restoration treatments, such as fire use and hand or mechanical treatments, to be restored to the historical fire regime.

Condition Class III:
Fire regimes have been significantly altered from their historical range. The risk of losing key ecosystem components is high. Fire frequencies have departed from historical frequencies by multiple return intervals. This results in dramatic changes to one or more of the following: fire size, intensity, severity, and landscape patters. Vegetation attributes have been significantly altered from their historical range.

Example of a management option: Where appropriate, these areas may need high levels of restoration, such as hand or mechanical treatments, before fire can be used to restore the historical fire regime.

So, this is where you need to put on your thinking cap to begin to understand how current condition classes overlay natural fire regime types.

There's an old adage that you have heard many times that one picture is worth a thousand words. Here's three pictures.

Figure 6-1: Photo taken in the 1880's in the Bitterroot Mountains of western Montana. Class 1 land.
Source: Healthy Forests Initiative

Figure 6-2: Notice how the trees and brushes have grown in the background. Source: Healthy Forests Initiative

Figure 6-3: A fire swept through the area in 2000.
Source: Healthy Forests Initiative

The three photos are a pictoral look of how forests grow up over the course of 100 years. One final note about the third picture. The house had been moved before the fire. Like an increasing number of fires, this fire took no prisoners.

Condition of national lands

The authors of the coarse-scale study assembled national data to develop a "ball-park" idea of the conditions of all of our lands (not just forestlands).

Of the 1.2 billion acres in our country, 47% of the lands are in Condition Class 1 - low departure, 38% are in Condition Class 2 - medium departure, and 15% are in Condition Class 3 - high departure (Table 6-2). Remember, we are mostly concerned about Classes 2 and 3 lands which total 53% of all lands (Figure 6-4).

Table 6-2: Condition Classes: All US ownership summary (a)
(thousands of acres)

	Class I		Class II		Class III			
Historical fire regime	Acres	%	Acres	%	Acres	%	Total acres	Total
I. 0 - 35 years; low severity	175,031	41%	176,161	41%	77,492	18%	428,685	34%
II. 0 - 35 years, stand replacement	192,544	57%	133,181	40%	10,346	3%	336,071	27%
III. 35 - 100+ years; mixed severity	127,642	43%	112,258	38%	54,002	18%	293,904	24%
IV. 35 - 100+ years; stand replacement	53,062	43%	35,333	29%	35,028	28%	123,424	10%
V. 200+ years; stand replacement	48,558	72%	13,706	20%	4,905	7%	67,170	5%
Total	596,839	47%	470,641	38%	181,775	15%	1,249,256	

(a) All ownership land summary of historical fire regimes by condition classes of all types except agriculture, water, and urban development

Source: US Forest Service - Development of Coarse-Scale Data for Wildland Fire and Fuel Management.

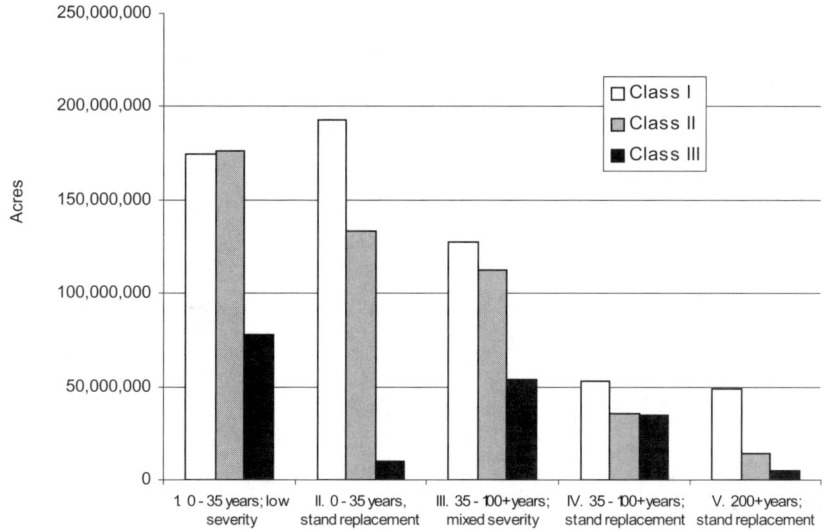

Figure 6-4: All US ownership fire regime condition classes

Excludes bodies of water and agricultural lands.

Source: US Forest Service - Development of Coarse-Scale Data for Wildland Fire and Fuel Management.

Condition of US Department of Interior lands

The US Department of Interior manages more public lands than any other federal land management agency. The vast majority of these lands are desert and high desert lands managed by the Bureau of Land Management. The Park Service lands, such as Yosemite National Park and Yellowstone National Park, are known as the "crown jewels" of our country.

The condition of the US Department of Interior lands is similar in proportion to the national condition of our lands. Summing up, 56% of Department of Interior lands are classified as Condition I - low departure - lands; 33% are classified as Condition II - moderate departure -, and finally, 11% are classified as Condition III - high departure. (Figure 6.5)

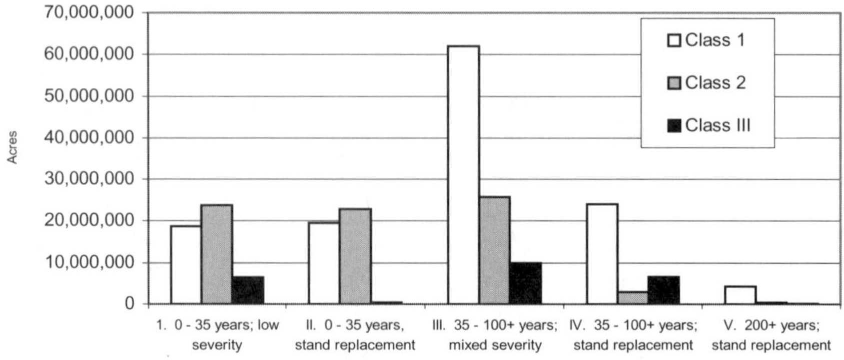

Figure 6.5: All Department of Interior Lands

Source: US Forest Service - Development of Coarse-Scale Data for Wildland Fire and Fuel Management.

Condition of US Forest Service lands

Although founded on the strong conservation ethics of President Teddy Roosevelt and Gifford Pinchot, the first Forest Service director, the national forest lands are suffering.

Looking at all Forest Service lands nationwide, 33% of the lands are classified in a normal range with low risk of fire modification, 40% of the lands are at a moderate risk, and 27% are at a high risk of modification from wildfires. Most importantly, 86% of the old growth stands of 200+ years are within the normal range (Table 6-3). The 200+ year old growth stands would tend to be located in wetter climates such as the Pacific Northwest and northern California. Nationally, two-thirds of all Forest Service lands are at moderate to high risk of modification from wildfires (Figure 6.6).

Table 6-3: Condition Classes: All US Forest Service lands
(thousands of acres)

	Class I		Class II		Class III			
Historical fire regime	Acres	%	Acres	%	Acres	%	Total acres	Total
1. 0 - 35 years; low severity	19,872	24%	34,961	42%	28,832	34%	83,667	43%
II. 0 - 35 years, stand replacement	4,458	33%	8,656	64%	360	3%	13,475	7%
III. 35 - 100+ years; mixed severity	16,046	30%	26,714	50%	11,165	21%	53,926	27%
IV. 35 - 100+ years; stand replacement	5,260	23%	7,352	32%	10,492	45%	23,105	12%
V. 200+ years; stand replacement	19,311	86%	2,760	12%	273	1%	22,345	11%
Total	64,949	33%	80,445	40%	51,124	27%	196,520	

Source: US Forest Service - Development of Coarse-Scale Data for Wildland Fire and Fuel Management.

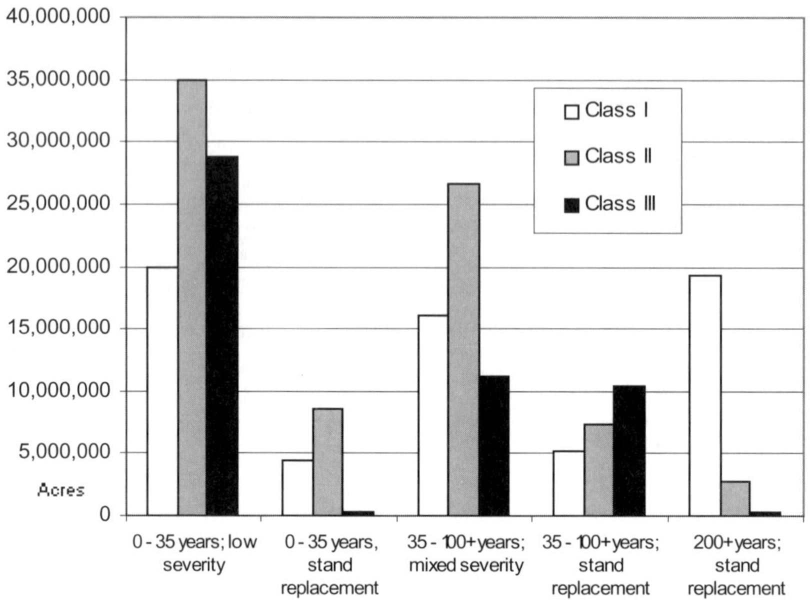

Figure 6-6: Condition of all US Forest Service lands

Source: US Forest Service - Development of Coarse-Scale Data for Wildland Fire and Fuel Management.

In Oregon and Washington, the situation is even worse for the percentage of lands that are classified as Class II or Class III.

Ironically, there is a greater percentage of national forest lands in Oregon and Washington in Condition II or III than nationwide. 78% of Forest Service lands in Oregon and Washington fall into either Class II or Class III lands (Table 6-4), of which one-third of the lands are categorized as Class III. The coarse scale maps of Oregon and Washington visually show the concentration of Class I, Class II, and Class III lands (Figures 6-7 and 6-9). The lightest gray color are non-federal Class I lands. The second lightest shade of gray are federal Class I lands. The moderate gray shade are Class II lands and the black shade lands are Class III (Figure 6.7 and 6.9). Figure 6.8 shows a chart of the acreages per fire regime classification for Oregon and Washington national forest lands.

Table 6-4: Condition of Forest Service lands in Oregon and Washington
(thousands of acres)

	Class I		Class II		Class III			
Historical fire regime	Acres	%	Acres	%	Acres	%	Total acres	Total
1. 0 - 35 years; low severity	78	1%	3,291	33%	9,972	66%	9,972	39%
II. 0 - 35 years, stand replace-ment	0	0%	48	32%	152	68%	152	1%
III. 35 - 100+ years; mixed severity	714	10%	5,036	70%	7,184	20%	7,184	28%
IV. 35 - 100+ years; stand replace-ment	914	30%	2,049	68%	3,028	2%	3,028	12%
V. 200+ years; stand replace-ment	3,758	70%	1,196	24%	7	0%	4,962	20%
Total	5,465	22%	11,623	46%	8,210	32%	25,299	

Source: US Forest Service - Development of Coarse-Scale Data for Wildland Fire and Fuel Management.

Figure 6.7 Condition Class 3 (Black) and Class 2 (Dark gray) Source: USFS

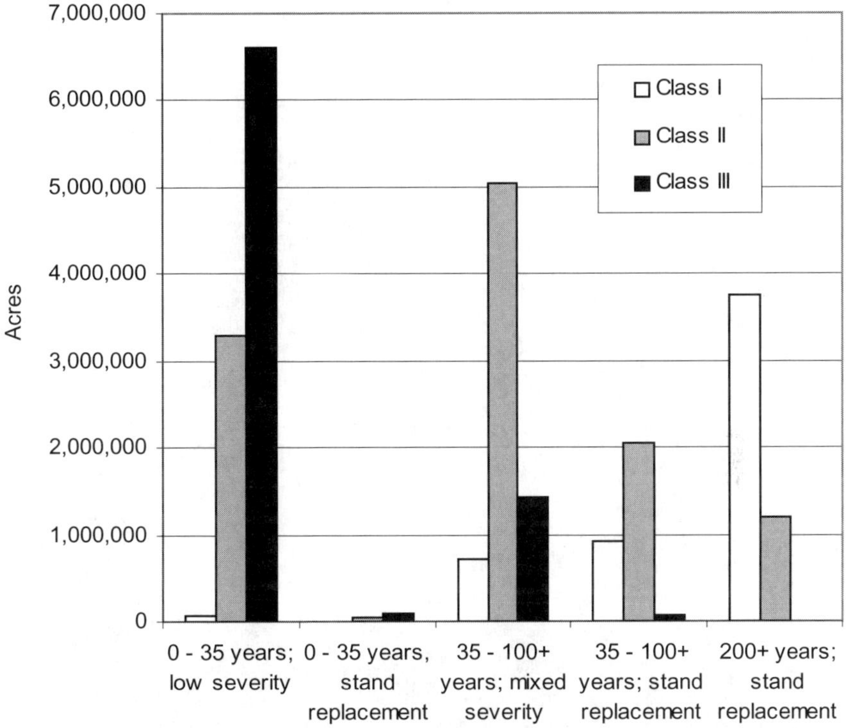

Figure 6.8: Condition of USFS lands in Oregon and Washington

Source: US Forest Service - Development of Coarse-Scale Data for Wildland Fire and Fuel Management.

I recognize that I have buried you with lots of numbers that are easily forgotten. Here's the primary point that I want to make. Our public and private forestlands in eastern Oregon, eastern Washington, and southwest Oregon are at extreme risk of wildfire destruction. These forests desperately need to be treated to reduce the dense, overcrowded stands.

Condition of private lands in eastern Oregon

According to the coarse scale study, private forestlands west of the Cascades tend to be in reasonably good condition

Figure 6.9 Condition Class 3 (Black) and Class 2 (Dark grey) Source: USFS

(with the exception of southwest Oregon). The picture is radically different for private forestlands east of the Cascades in both states. Private forestlands mirror much of the same problems facing public lands east of the Cascades.

A 1994 study entitled *Timber resource statistics for timberland outside national forests in Eastern Oregon* by N. McKay, M. Mei, and Gary Lettman showed that softwood mortality was nearly 35% of the gross growth. By comparison, the softwood mortality for the entire Pacific Coast region (including Washington, California, Oregon, and Alaska) averaged only 7%. The study showed that 29% and 40% respectively of all live trees on industry lands and non-industrial forestlands showed signs of an insect attack. Most of the affected trees were either lodgepole pine, Douglas-fir, white fir, and grand fire.

A 2003 Washington DNR survey found that 1.9 million acres of land had increased levels of tree mortality, tree defoliation, or foliar disease. Almost three million trees were recently killed, an increase from 1.84 million killed in 2002.

What needs to be done

The *Fire and Land Management Planning and Implementation Across Multiple Scales* report, by David Bunnell and Wendell Hann, examines the state of ecosystem conditions on all federal lands in the United States. The authors summarized numerous studies about the decline of the ecosystems. Bunnell and Hann found that the cumulative area burned by wildfires in the western United States decreased for forty years after the large fire season of 1910, which prompted the beginnings of widespread federal fire suppression efforts. The authors found that wildfires have steadily returned to levels comparable or greater than those encountered at the beginning to the 20th century.

Bunnell and Hann created three modeling scenarios to simulate future outcomes:
1) Continue current treatments at 0.7% per year – status

quo in terms of what the authors perceived as to the level of management. That is, that .7% of all Forest Service lands were annually treated.

2) Increase current treatments to 2% per year – Assumes that the lands will be managed as in the continue current scenario, but with an increase up to 2%.

3) Integrated at 2.7% per year – assumes that desired outcomes are integrated within appropriate landscape scale of delineation.

The first model assumes that there would be approximately 2 million acres treated per year while the third model projects 3.7 to 4.5 million acres treated annually.

In plain language, the Forest Service authors ran a number of computer models both looking back at the historical data for the past 100 years and looking forward for the next 100 years. The computer models showed that that there would be a continuing increase in the percent of land burned by wildfires in the next century under the current treatment level of .7% but that there would be a significant decrease in wildfire activity with a proposed "integrated at 2.7%" scenario.

What has happened with fuels reduction funds

Several years ago, I analyzed the financial aspects of the *Hazardous Fuel Reduction on Federal Lands for the Pacific Northwest* brochure. Congress appropriated $242.5 million to the BLM and USFS in Oregon and Washington for FY 2001. The USFS received $184.7 million from the National Fire Plan while the BLM received $57.8 million. The agencies used $25.1 million of these funds to contract for goods and services. Of that $25.1 million, the USFS procured $10.3 million and the BLM procured $14.8 million through contracting. I have combined the Oregon and Washington totals separately to calculate the per acre cost of the treatment.

Table 6-5: Hazardous Fuels Reduction in Oregon					
Agency	**USFS**	**BLM**	**FWS**	**NPS**	**BIA**
Prescribed fire (acres)	43,309	35,704	4,378	1,300	965
Mechanical treatment	33,505	20,835	52	15	29
Total acres	76,814	56,539	4,430	1,315	994
Total cost (millions of dollars)	$15.0	$20.7	$.36	$.2	$1.5
Cost per acre	$195	$367	$81	$214	$1,565
Source: Personal analysis					

Table 6-6: Hazardous Fuels Reduction in Washington					
Agency	**USFS**	**BLM**	**FWS**	**NPS**	**BIA**
Prescribed fire	20,270	750	9,936	52	4,854
Mechanical treatment	5,019		326	318	250
Total acres	25,289	750	10,262	370	5,104
Total cost	$3.4		$1.3	$.1	$1.2
Cost per acre	$137	-	$128	$267	$244
Source: Personal analysis					

Based upon my limited knowledge of bidding thinning contracts and losing thousands of dollars, the per acre treatments are quite low. Manual brush disposal can run anywhere from $500 to $2,000 per acre in southwest Oregon. Prescribed burning costs may run anywhere from $25 to $500 per acre - depending upon the vegetation, topography, and weather. Adjacent units may have radically different costs because of weather conditions.

The one major factor that will determine the treatment cost per acre is the willingness of the contractor to use illegal aliens (undocumented workers) to perform the work. Although I will be chastized for being politically incorrect and insensitive, white lads and lasses can not perform the work as productively as male Hispanics. (I might note that the private industry is just as willing to hire illegal aliens to perform the work).

The Forest Service and BLM spent $39,249,000 (which presumably includes both internal as well as contracting crews). How was the remainder of the $242.5 million spent?

The bottom line is that there are not enough dollars that are getting down to the physical treatment of the lands. Despite the agencies' inability to get money to the ground, I am a strong proponent of rebuilding both the Forest Service and BLM agencies. Bottom-up leadership is needed to insist that the majority of the funds are spent on the ground.

Quickie

The key point in this long series of statistics is that the forestlands are suffering from the growth of too many little trees and too much brush. Of the 25.2 million acres of Forest Service lands in Oregon and Washington, 46% of the lands are classified as Condition 2 moderate risk and 32% are classified as Condition 3 high risk. The Kalmiopsis area in southwest Oregon was largely classified as Class III high risk lands in 2000. Two years later, the Biscuit Fire ripped and burned through 500,000 acres in the Kalmiopsis.

Chapter 7

The threat of wildfires

For an example of how times have changed, visit

www.smokeybear.com

Whipped by the hot, dry easterly Santa Ana winds, a series of wildfires ripped through Southern California in late October 2003. In five brutal days, the fires devoured 750,000 acres, killed 20 people, burned 3,500 structures, and caused over $2 billion of property damage.

I traveled to southern California in early January 2004 for a post-Christmas get-together with my father and brother in San Diego. I drove up to Cuyamaca State Park, 35 miles east of San Diego. Cuyamaca Peak is the highest point in San Diego County with an elevation at 6,500 feet. Unlike the surrounding chaparral-covered hills of San Diego County, Cuyamaca Peak was graced with Jeffrey pines, cedars, and California white oak. I had heard that one of the October 2003 fires had burned through the park, but I was not prepared for the devastation. In my many years working on Forest Service and Park Service hot shot crew and fire suppression crews, I had never seen such a devastating fire. Whereas most fires skip over patches of vegetation and burn through the under story of trees, the fire roared through the precious forest stands. There was nothing but black sticks for as far as the eye could see. An analysis by California state employees showed that only 300 of the 25,000 acres of park forests survived the carnage.

What is most sad about the park is the realization that the cash-strapped state of California will simply turn its back on the state park and never reforest the park. Rather, nature will be allowed to reseed the area. The problem is that there is little forest seed stock in the immediate area. In the next twenty to thirty years, the peak will simply become another chaparral-covered mountain in San Diego County.

In this chapter

Fires are a natural part of the ecosystem. Fires are beneficial to Mother Nature and play a very key role in re-starting the ecological processes. However, many of the fires burning today can not be characterized as Mother Nature's natural processes. Fed by unnatural increases in tree and brush growth as well as the global warming wild card, these fires are roaring out of the chute. This chapter will discuss the very real threat of wildfires to destroy a significant amount of our ecosystem in the Pacific Northwest.

The initial wake-up call in 2000

The 2000 fire season was the first wake-up call to Americans lulled by years of wildfire inactivity. The 2000 fire season was the worst fire season in fifty years. Nearly 120,000 fires nationwide burned over 8,000,000 acres of fires in 2000. The ferocity of the 2000 fire season prompted the Department of Agriculture and Department of Interior to prepare a report entitled: *Managing the Impacts of Wildfires on Communities and the Environment.* State governors, other partners, and federal agencies joined together to develop a 10-year comprehensive strategy to address the problem. 2001 was an average fire season with some false expectations that the nation had somehow gotten prepared.

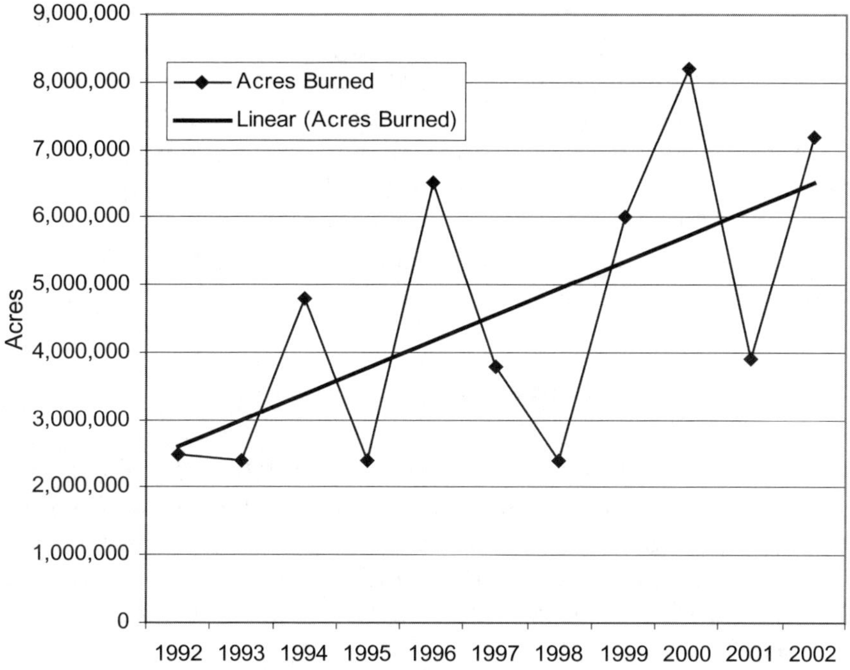

Figure 7-1: Wildland fires 1992 - 2002

Source: National Fire Plan

If the 2000 fire season was an initial wake up call, the 2002 fire season was a shot across the bow that there were serious problems that would not be wished or organized away (Figure 7.1). 2002 was the second largest fire season in fifty years. A widespread drought – reputed to be the worse since the Dust Bowl of the 1930s - affected 45% of the country's land base. Three states recorded their largest timber fires in the last 100 years: Arizona, Colorado, and Oregon.

At the peak of the 2002 fire season, the federal and state governments had 28,000 people, 1,205 fire engines, 30 air tankers, and 188 helicopters assigned to fires on July 31. By the end of the fire season, 73,000 fires had burned 7.2 million acres of land, or approximately twice the 10-year average.

The Biscuit Fire gets an afternoon run

On July 13th, 2002, a small lightning fire started in Florence Creek in the Kalmiopsis Mountains of Siskiyou National Forest. Due to its remoteness, adjacent to the Kalmiopsis Wilderness, and unavailability of resources which had been committed to other lightning fires, the Biscuit fire steadily grew. The Biscuit Fire burned nearly 500,000 acres for two months before cool weather allowed the federal agencies to declare the fire contained.

The Biscuit Fire grew so big that the fire got its own web address: www.biscuitfire.com. According to the Biscuit Fire web site, *"The fire burned in a mosaic pattern; approximately 20% of the area burned lightly, with less than 25% of the vegetation killed. Another 50% of the area burned very hot, with more than 75% of the vegetation killed."* Many acres of critical habitat for wildlife burned, and the late seral and old growth stands that remain are very precious.

The fire burned from July 13th to November 19th for a total of 499,965 acres. Over 7,000 firefighters and support personnel helped to "contain" the fire at its peak. The total cost of the fire to date is $153 million. The Biscuit Fire burned the entire Kalmiopsis Wilderness.

Fires are expensive

When I worked for the Forest Service and National Park Service on fire crews, we would have fire camp at the beginning of each summer to review fire basics. We reviewed the fire triangle concept: fuel, heat, and oxygen. If you remove any of these three elements from the fire, the fire would go out. For example you could throw dirt on the fire to smother it. All of us knew that was not the real truth. The real truth was the fire square: fuel, heat, oxygen, and overtime. If you removed any of those four elements, the fire would go out.

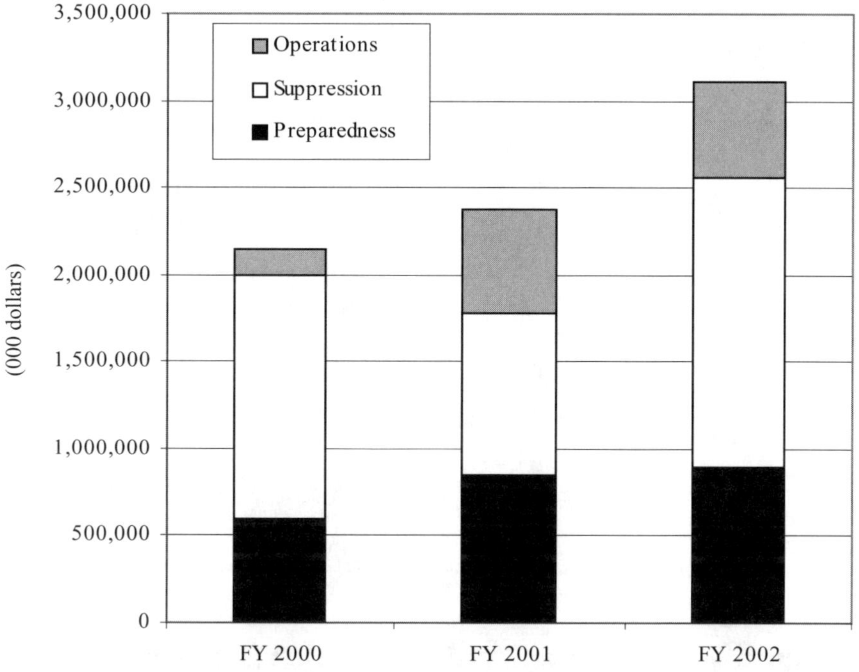

Figure 7-2: Wildland fire appropriations for FY
Source: FY 2002 Performance Report: National Fire Plan

Like all federal and state agencies, the Forest Service and BLM need to budget for the next fiscal year. Budgeting for the amount of wildfire activity is difficult if not impossible to predict since there are wide variations. For example, the Farmer's Almanac is not necessarily a guarantee or good prediction of what will happen the following summer.

The budget activities of the Forest Service and BLM include four major categories:

Preparedness:	Getting crews and equipment ready for the fire season.
Operations:	Basic activities
Emergency:	Set aside funds
Suppression:	Actual suppression costs

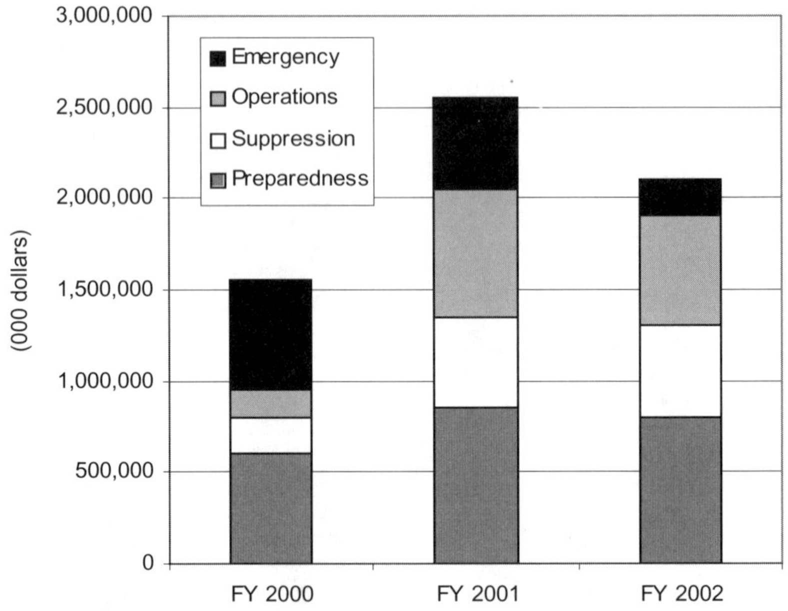

Figure 7-3: Actual Wildland Fire Expenditures
Source: FY 2002 Performance Report: National Fire Plan

The 2000 and 2002 fire seasons burned a big hole in the pockets of the US Forest Service and Department of Interior. The 2000 and 2002 fire seasons cost $1.5 billion and $1.666 billion respectively.

Actual fire suppression costs are the biggest expense category. Furthermore, suppression funds – because of the hurried nature in which these funds are expended – are not efficiently spent. A governmental agency may casually spend $1 million on a single fire on a sunny August afternoon.

National Fire Plan expenditures

The FY 2002 Performance Report: National Fire Plan provides a state - by - state breakdown of how funds were expended. Of the $3.18 billion expended nationally, $2.4 billion was spent in the western states (Figure 7-4).

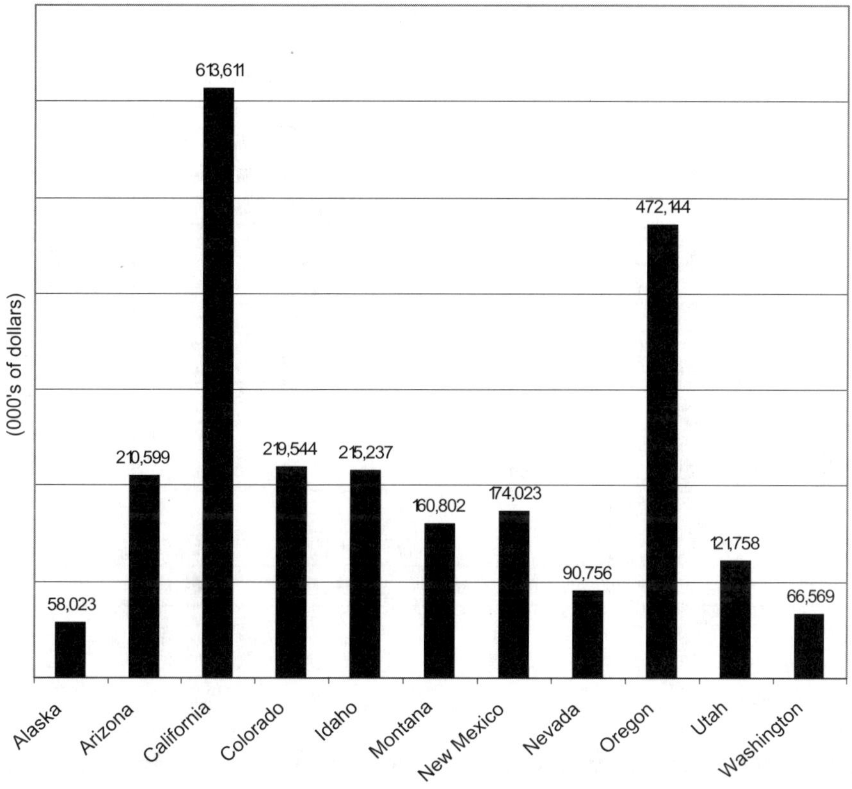

Figure 7-4: Fire spending by state
Source: FY 2002 Performance Report: National Fire Plan

It is understandable that two thirds of the $3.1 billion funds were expended in the ten western states, since the federal government is the largest landowner of the West.

The amount of funds paid by the government for contractors reveals the priority of the federal government. This table display which types of services and products were purchased by the federal government in FY 2002 for natural resources management such as fuels treatment. That is,

Table 7-1: FY '02 National Fire Plan Contracting by Program

(Displayed in millions of dollars)

Program	Fuels Treat-ment	Prepared-ness and facilities	Rehabili-tation	Suppres-sion	Total
BLM	$13.2	$3.1	$13.8	$6.5	$39.8
FWS	$0.4	$4.2	$1.6	$3.2	$9.6
NPS	$0.1	$0.9	$-	$1.6	$2.7
BIA	$14.2	$12.4	$2.2	$7.6	$36.5
DOI total	$28.1	$20.8	$17.7	$19.1	$85.8
FS	$10.5	$18.0	$13.1	$201.9	$243.6
DOI/FS total	$38.6	$38.9	$30.9	$221.0	$329.5

Source: FY 2002 Performance Report: National Fire Plan

Table 7-1 shows what the government used the private sector to perform which programs. The federal government spent $243 million of its $329 million of contract dollars for fire suppression costs (Table 7-1).

The problem is that many contractors have shifted from reforesting, thinning, and restoring the forests to hoping that there are big fires so that they can make money from the calamities. While local residents afflicted by wildfires may put up signs to thank the firefighters, the primary reason that the firefighters are there is to make money off the fire. If there was no money, it's hard to imagine that there would be thousands of Americans volunteering their time to put out these fires.

As Senator Dirksen once said forty years ago, "a million here, a million there, and pretty soon you're talking real money."

Quickie

The 2000 fire season was the worse fire season in fifty years while the 2002 fire season was the second worse fire season. Fire suppression costs exceeded $1.5 billion in FY 2000 and FY 2002. Firefighting can be a very inefficient and wasteful expense of money. The Biscuit Fire burned 500,000 acres in southwest Oregon in 2002 which was the biggest fire in the last 100 years in the state. The number of acres burned and the fire suppression costs will continue to increase.

Chapter 8

How to use laws, policies and grants to create jobs

You've seen the bumper sticker: Stumps don't lie. That's not the whole story. Stumps don't tell the truth, either.

After graduating from the Lundquist College of Business at the University of Oregon in June 2003, I spent many hard, fruitless months looking for work. Over the next several months, I sent out over 250 cover letter / resumes for advertised jobs and filled out over 100 job applications. While I was logged into the Oregon Employment Department computers searching fruitlessly for financial and accounting work, I clicked on the forestry occupation out of curiosity. While the financial, business, technical occupation would typically have 400 – 500 job listings for the state, there were only 85 forestry jobs displayed for the entire state of Oregon. How pathetic. The state of Oregon has more softwood volume on its lands than any other state in the country. I had one of those "crystal clear" moments when you realize that something is "really rotten in Denmark."

In this chapter

This chapter will focus on how new laws such as the Healthy Forests Restoration Act and new authorities such as stewardship contracting can be used to create jobs.

Just as environmentalists have used laws under the belief that they are protecting the forest ecosystem, a new set of laws and policies can similarly be used to create jobs on our public and private forestlands and actively protect the environment from wildfire ravages. We can create forest jobs in rural and urban areas that can compete on the open, global market.

What types and how many jobs can be created? I'll be covering those questions in Chapter 12 with a number of models. Are those models and figures reliable? No.

What I can do is to show you that there are a variety of different types of jobs that can be created because of changes in law. More importantly, there have also been changes in public opinion. According to a poll conducted by the Oregon Forest Resources Institute in 2003, 85% of Oregonians recognize that fires represent a serious danger in the state.

Healthy Forests Restoration Act

Termed the most important forest legislation enacted in the past 25 years, the Healthy Forests Restoration Act of 2003 (HFRA) sailed through US Senate on a bi-partisan vote of 80-17 in the aftermath of the disastrous October 2003 fires in California that seared 750,000 acres, killed 20 people, destroyed 3,500 structures and caused $2 billion of property damage.

Regardless of how you may feel about President Bush, the Healthy Forests Restoration Act makes good environmental and economic sense. The common sense and pragmatic approach of the HFRA suggests that President Bush, Vice President Dick Cheney, Undersecretary of Defense Paul Wolfowitz, Defense Secretary Donald Rumsfield, and National Security Advisor Condeliza Rice were too busy on other worldly projects at the time to screw up the Healthy Forests Restoration Act. Indeed, the act appears to have been written by a pragmatic group of professionals who were focused on the actual problems rather than on making ideological statements that marks much of the Bush Administration. Environmentalists should be embracing the act because of its ability to transform our sick forests. However, it does not fit their agenda for extremism and does not give them control.

The Health Forests Restoration Act suggests that we are likely entering a paradigm shift from the "analysis paralysis" that has gripped the Forest Service and BLM for the past ten years. Ironically, the "analysis paralysis" has not been good for the forests nor will the paralysis have been good for environmentalists in the long run. It has been good for certain environmentalists because their goal is to control everything with a simplistic, preservationist, unscientific approach. Some environmentalists have not shown much interest in generating tax review or creating jobs. "Benign neglect," as paraphrased by Senator Moynihan many years ago, has not been good for our forests.

What does the Act do?

The new forestry act includes a number of key provisions:

1) Categorical exclusions of 1,000 acres for forestry restoration projects.

2) Limits appeals to parties involved in the development of the projects.

3) Categorical exclusions for up to 4,500 acres of prescribed burning.

4) Limits appeals in fire restoration projects.

5) Allows limited timber sale harvesting up to 70 acres.

Jim Riley, president of the Intermountain Forest Association, spoke at the Small Log Conference in April 2004 about the change that is occurring at the Forest

Service. He cited figures showing that the Lolo National Forest had gone from zero categorical exclusion projects in FY 2003 to 32 categorical exclusion projects treating 4,100 acres in FY 2004. The Panhandle National Forest had also jumped from zero categorical projects in FY 2003 to 38 categorical exclusion projects in FY 2004 treating over 7,000 acres. He stated, "I view this as encouraging information."

Direct jobs to be created

Logging
Trucking
Mill workers
Secondary wood processing
Furniture makers
Public sector jobs
Surveyors
Monitors
Forestry service workers
Pre-commercial thinners
Office staff for each of the
above job classifications

Indirect jobs to be created

Retail jobs
Wholesale jobs
Transportation jobs
Information jobs
Other government jobs
Hospitality and leisure

He noted that the key will be for the federal government to carry out landscape projects and to provide a "continued and sustainable timber supply."

How the Act will create jobs

The Healthy Forests Restoration Act will begin to create a new set of jobs once the private sector sees that the Forest Service and BLM are serious about addressing the over-crowding of the forest stands. If the Forest Service and BLM offer up a stable and predictable supply of small log timber sales, the private sector will begin to respond.

The Act has already helped the Forest Service win more cases in court against some environmental groups seeking to stop active management. The Forest Service won 14 straight cases in March 2004. Mark Rey, the Agriculture Department undersecretary, said, "Any month when you win 14 out of 14 cases is a good month. Judges are required by the new act to balance the environmental consequences of no action and the risk of wildfire when considering tree cutting projects. Further, judges must review any court orders blocking projects every 60 days.

Stewardship Contracting

The Forest Service and BLM used to operate with two major types of contracting tools: timber sales and service contracts. A company would bid and pay the Forest Service for removal of valuable timber products. The contractor might be responsible not just for the removal of timber, but for site preparation such as piling of the brush. The contractor might also be responsible for building roads into the forest. As noted in Chapter 5, the Forest Service and BLM made billions of dollars from timber sales.

Stewardship contracting is a hybrid cross between timber sale contracts and service contracts. Stewardship contracting was approved in the 2003 Appropriations Act (P.L. 108-71). According to a January 15, 2004 joint news release of the US Department of Agriculture and US Department of Interior, stewardship contracting will help federal agencies achieve key land management goals to:

1) Improve, maintain, and restore forest and rangeland health;
2) Restore and maintain water quality;
3) Improve fish and wildlife habitat;
4) Reestablish native plant species and increase their resilience to insects, disease, and other natural disturbances;
5) Reduce hazardous fuels that pose risks to communities and ecosystem values through an open, collaborative process.

Stewardship contracts will allow private companies and communities to retain forest products in exchange for the service of thinning trees and brush and removing dead wood. The government hopes that stewardship contracts will foster a public / private partnership to restore public forestlands by giving those who undertake the contract the ability to invest in equipment and infrastructure. The government allows community groups and private companies to enter into stewardship contracts for up to 10 years. Contracts are awarded on "best value" which allows for consideration of performance as well as bid price. The language also allows multiparty monitoring to assess the projects. One local benefit of stewardship contracting allows for some of the revenues to be kept on the forest so that the funds may be reinvested in other stewardship projects. For those who are interested in more rule details, you may review the Forest Service Handbook 2409.19 – Renewable Resources Handbook, Chapter 60 – Stewardship Contracting.

"Stewardship contracting will demonstrate a 'new environmentalism' – land stewardship based on partnerships and common ground rather than litigation and confrontation," said Rebecca Watson, DOI assistant secretary. "It is part of a new culture of communication, cooperation, and consultation, in the service of conservation – a culture that Secretary Norton calls the "Four C's."

Both departments released a 57-page Interim Field Guide to help government employees understand the new regulations. The field guide is currently located at:

www.nifc.gov/stewardship.html.

Congress first approved pilot stewardship projects in 1999. The agencies have completed 70 pilot projects to date. As with any new program, there have been both successes and problems with the pilot stewardship projects. An excellent monitoring report by the Ecosystem Workforce Project and the Watershed Center about Oregon and Washington stewardship projects is located at:

http://ewp.uoregon.edu or www.thewatershedcenter.org.

Federal officials are placing great importance on stewardship contracting. While the agencies recognize that stewardship contracts are simply one more tool for forest management, the agencies are hoping that stewardship projects will be less susceptible to appeals and will receive broader public acceptance.

Stewardship contracting may encourage local entrepreneurs to approach the Forest Service and BLM with proposals to manage small sections of land. For example, a contractor or community group may propose to thin out several hundred acres of small logs in a flat area. Part of the proceeds from the exchange would be to require the contractor to perform some type of watershed enhancement project such as placing some of the logs in nearby creeks. Similarly, the contractor or community group may propose that some of the small logs be ground into chips and those chips then laid into a trail area that suffers from too much wet soil.

How to use government grants to create jobs

The National Fire Plan has three key grant programs to help out communities and rural fire departments in the Pacific Northwest:

1) Volunteer and Rural Fire Department Assistance
2) Community Assistance and Economic Action Programs
3) State Fire Assistance Wildland Urban Interface Hazard Mitigation Grants

Volunteer and Rural Fire Department Assistance – This program is directed towards providing financial assistance to rural fire departments in the country. Rural fire departments in Oregon and Washington are suffering from a lack of volunteers and lack of necessary equipment. The grant funds may be used for training, organizing, and equipping fire districts. The funds may not be used for land acquisition or major capital projects. Total award amounts are approximately $1.2 million in Oregon and $680,000 in Washington. Grants are administered by the Oregon Department of Forestry and the Washington Department of Natural Resources. The grantee must offer to provide 10% to 50% in matching money or in-kind services. Washington and Oregon grant applications are due by June 15th and April 10th respectively. Further information may be obtained from Washington at ***www.wa.gov/dnr/base/assistance.html*** and Oregon by e-mailing to the agency at ***dmatlick@odf.state.or.us.***

Another significant grant aid program for rural fire departments is administered by the US Fire Administration of the Federal Emergency Management Agency. This agency awarded $8.7 million to Oregon rural fire departments in 2003. The national program awarded 8,100 fire departments almost $651 million in 2003.

Community Assistance and Economic Action Programs -

This combined grant program of the US Forest Service and BLM distributes approximately $7 to $9 million annually in Oregon and Washington. The total amount to be awarded by agency is $3.2 million USFS, $3 million BLM, $2 million FWS, $1.3 million BIA, and $40,000 National Park Service.

The four combined programs under this grant are:

1) Wildland Urban Interface fuels treatment projects;
2) Wildland Urban Interface Community Risk Assessment and Mitigation Planning;
3) Wildland Urban Interface Prevention and Education projects;
4) Wildland Urban Interface Fuels Utilization and Marketing.

The funds may be used for community fire planning, education, developing business plans, marketing tools, training and technical assistance, purchasing wood processing equipment, and fuels hazard reduction ***(www.fs.fed.us/r6/coop).***

The grant funds are limited to local governments, Indian tribes, universities and non-profit organizations. Grantees are required to match 20% of non-federal money or in-kind services. The application date is February ***(www.fs.fed.us/r6/coop).***

State Fire Assistance Wildland Urban Interface Hazard Mitigation Grants – These grants are to be used to reduce the threat of fires in the Wildland urban interface among the 17 western states. The funds may be used for hazard mitigation, fuels and risk reduction, information and education programs for homeowners and communities. Only state forestry agencies may apply for these programs. The state forestry agencies, in turn, will award fuels reduction work ***(www.fs.fed.us/r4/sfa_grants/sfa_grants.html).***

Quickie

The Healthy Forest Restoration Act is the most significant forestry legislation act in the past 25 years. The Act has the potential to jumpstart active forest management on our public lands to reduce the excessive fuel loading. Representative Peter DeFazio (D-OR) introduced legislation in August 2004 to encourage thinning of government forestlands west of the Cascades. A number of federal fire grant programs are available.

Chapter 9

How to pay for transforming our forests

While the growing need for sustainable electric power can be met by other renewables . . .

Biomass is the only renewable that can meet our demand for carbon-based liquid fuels and chemicals

National Renewable Energy Laboratory

I have to confess that I am an addict. I was sitting at the Beanery in downtown Eugene getting my daily latte fix when I saw two county workers roll up a cart to the Lane County Adult Corrections Jail and Mental Health Hospital wooden sign across the street. I saw one of the workers pull out a large wooden sign from the cart. I figured that there was a name change for one of the two facilities. Instead, the new wood sign was blank. The workers screwed the blank sign over the Mental Health Hospital name and rolled the cart away. I remembered reading in the newspaper the previous week that Lane County was closing the Mental Health Hospital due to a lack of funds.

In this chapter

This chapter will describe three broad strategies to show how we can treat these lands and eventually save on the expenditure of tax dollars for fire suppression. Two of these strategies have already been implemented in parts of the country while the other strategy has both national and international implications. The three economic strategies are:

1) Small log utilization
2) Biomass conversion into energy
3) Ethanol production

The federal government is going broke

If the Bush Administration was running a surplus budget or not running an expensive war(s) in the Middle East, there would not be a question that would need to be posed as to how to pay for the tens of millions of acres of forestlands that need to be treated on public lands. However, since the Bush Administration ran up a $380 billion dollar deficit in FY 2003 and with red ink flowing into the future as far as the Congressional Budget Office is willing to look, treatment strategies need to be developed that will cost the least to the taxpayer.

The good news is that we can have part of our cake (protecting the forest environment) and eat part of it too (paying for the treatment without using too many tax dollars). The acre cost per treatment varies quite a bit depending upon the topography and brush species. My own personal experience suggests that the fuels reduction costs will vary anywhere from $400 to $2000 per acre. The trick will be to figure out how to generate enough money from the forest products to pay for the treatment costs.

Fuels reduction work requires a long-term stream of income to pay to perform that work. The income can come from either the public or private sector. The Bush Administration has proposed approximately $740 million for fuels reduction work for FY 2005. Private sector funding is better in the long run because of the market economy of the U.S. and the crushing public debt. The more streams of dollars being generated from different operations will likely increase the number of jobs that can be created.

What is the economic value of a tree?

For readers who are unfamiliar with commercial utilization of a harvested tree, only 28% of a typical tree is available for sawn timber, e.g., 2 x 4's, 2 x 6's (Table 9-1).

Table 9-1: Division of a typical tree harvested for sawntimber

Tree part or product	Portion (%)
Left in the forest:	
Tops, branches and foliage	23.0
Stump, (excluding roots)	10.0
Sawdust	5.0
Sawmilling:	
Slabs, edgings and off-cuts	17.0
Sawdust and fines	7.5
Various losses	4.0
Bark	5.5
Sawn timber	28.0
Total	100.0

Source: www.fao.org/docrep/to269e/t0269e08.htm

Fuels reduction work will primarily focus on small-diameter trees rather than large-diameter trees. The market value for small-diameter trees converted into poles is approximately $235 per green ton (Figure 9.1). Posts have an approximate value of $220 per green ton while chips have an approximate value of $25 per green ton. (One timber buyer told me that chips do not have any value and need to be delivered to the mill at zero cost value). A green ton (GT) weighs approximately one and a half times as a bone dry ton (BDT). The last column show the harvest cost of the trees which is approximately $75 per ton (Figure 9.1)

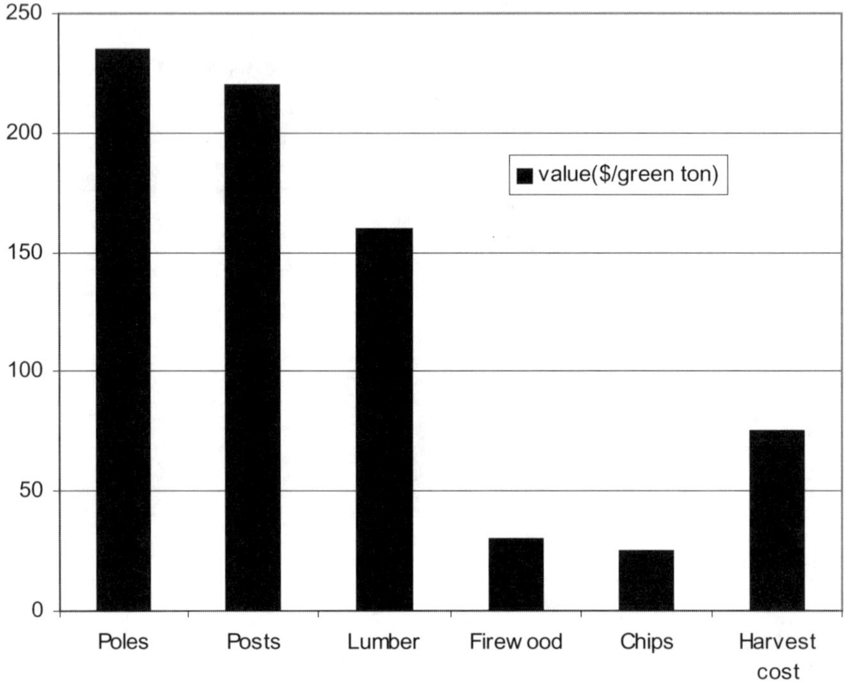

Figure 9.1: Market value for small-diameter trees.
Source: Healthy Forests, Thriving Communities: BioEnergy and Wood Products Conference

How do the market values relate to the real world.

A company that wishes to harvest a tract of land must consider the amount of the trees that can be harvested as well as the types of products that can be created from the trees. The extraction costs of a big tree vs. a little tree are similar, but the volume and value of a big tree is much greater. You have to go out, cut the tree, and haul the tree from the forest back to the mill for processing.

There are three main categories for the classification of harvested trees in terms of their value:

1) Sawlogs - greater than 11 ½" diameter
2) Small logs - 4 ½" – 11 ½" diameter
3) Pulp logs - less than 2 ½" diameter or cull / dead

How much biomass is available

How much "excess" biomass is available from our public lands in Oregon and Washington? Bryce Stokes, National Program Leader for the USDA Forest Service gave a presentation at the *Healthy Landscapes, Thriving Communities: BioEnergy and Wood Products Conference* in January 2004 about the national amount of biomass available. I'm sure that you remember the 190 million acres of land figure that is in need of fuels reduction treatment. He estimated that if you were to just focus on removing excess biomass volume from Class 2 and Class 3 lands, you would need to harvest nearly 40 times the annual conventional amount.

Bill Carlson, vice president of Business Development for Wheelabrator Technologies, estimated that the government needed to remove 380 million tons per year which would be equivalent to one half of the annual output of the US coal industry at the *Healthy Landscapes* conference.

Duane Vaagen, president of Vaagen Bros., approached the issue from a different point and came up with figures that were approximately twice the amount. Vaagen provided these estimates at TimberWest's Small Log Utilization conference in Coeur d'Alene, Idaho in April 2004.

190 million acres of priority treatment

190 million acres at 5,000 board feet /acre = 950 billion board ft

80 year treatment cycle = 11.9 billion board feet annually

The Forest Service, in partnership with the Western Forest Leadership Coalition, produced a report entitled *A Strategic Assessment of Forest BioMass and Fuel Reduction Treatment in Western States* in April 2003. The authors of the report estimated 1.493 million bone dry tons of Class 2 and Class 3 volume to remove from the 11 Western states (Table 9-2)

Table 9-2 Standing and removal volume by state (million bone dry tons)

State	Treatable timberland	Volume from Class 2 & 3	Class 3
AZ	61.3	57.8	39.5
CA	368.1	317.9	125.1
CO	165.3	99.9	48.3
ID	242.9	177.6	80.4
KS	21.9	6.1	2.5
MT	271.3	188.2	77.4
NE	11.5	2.1	.5
NV	2.8	1.7	.5
NM	57.5	49.6	32.2
ND	5.5	.3	0
OR	436.6	291.1	91.0
SD	11.1	9.0	6.6
UT	54.5	15.6	1.6
WA	371.5	241.5	63.0
WY	72.1	34.7	7.5
Total	2154.0	1493.4	576.1

Source: *A Strategic Assessment of Forest BioMass and Fuel Reduction Treatment in Western States,* April 2003

The top three states are California, Oregon, and Washington, in terms of the volume to remove (Table 9-2). The three states account for approximately 56% of all the treatable volume to be removed.

The biomass report estimated and analyzed the "generic" cost of fuels treatment and the trade-offs (Table 9-3):

Table 9-3: Generic comparison of fuel reduction treatment alternatives

Treatment	Cost Range	Key Benefit	Key problem	Products?
Prescribed fire	$35 - $300/ac	Low cost	Restricted use	No
Mastication	$100 - $1000/ ac	No smoke	Fiber left in woods	No
Cut/pile/burn	$100 - 750/ac	Low access	Burning limitations	No
Cut/skid	$30 - $40/ bdt	Offset costs	Soil impacts	Yes
Cut/skid/chip	$34 - 48 / bdt	Usable fiber	High cost, low value	Yes

Source: *A Strategic Assessment of Forest BioMass and Fuel Reduction Treatment in Western States,* April 2003

There is no generic treatment option for these lands. Each of the areas that need to be treated need to be considered on a case-by-case basis.

Small log utilization

There are a number of small log mills that have opened in the last few years to take advantage of the "wall of small wood" that is beginning to overwhelm the market. Until current technology improved, small logs were either left on the unit to be burned or rot away. This chapter highlights three companies that are successfully processing small logs. Two of the firms are owned by tribal nations and the other firm is privately owned.

Confederated Tribes of Warm Springs

The Confederated Tribes of Warm Springs reservation is located in central Oregon, approximately 65 miles northwest of Bend. Much of the tribal lands abut Forest Service lands. Virtually all of the Forest Service lands are classified as Current Condition Class 3 lands at risk of high fire danger.

The Deschutes National Forest suffered a 50,000 acre B & B Complex fire in August 2003 that burned primarily through Class 3 lands in close proximity to Warm Springs' boundary.

The Confederated Tribes operates a for-profit sawmill that provides employment for 132 tribal members and supports 101 tribal families. The Warm Springs Forest Products Industries manages a "two sided" mill which includes a traditional mill that processes large logs in the range of 18" to 52" as well as a small log facility that processes logs in the range of 8" to 17". The mill anticipates that 25% of its future raw materials will fall into the 5" to 7" diameter class. The reservation forest currently provides 90% of the sawmill's needs, but the Warm Springs Forest Products Industries is currently looking at the surrounding national forests for additional raw material.

Larry Potts, WSFPI general manager, spoke at the Small Log Utilization Conference sponsored by Timber West, Coeur d'Alene, Idaho, in April 2004 about the accomplishments of the company. Mindful that the mill must interact with the community at large, Potts described how WSFPI has become involved with the local planning groups such as the Central Oregon Intergovernmental Council, the Central Oregon Partnership for Wildfire Risk Reduction, the Central Oregon Renewable Energy Economic Development, and the Coordinated Resource Offering Protocol.

The mill is continuing to add value to its current operations by receiving environmental certification from the Forest Stewardship Council. The mill has proposed building a 15 megawatt biomass energy facility for PacificCorp.

Yakama Nation

A second example of a successful small log sawmill is located on the Yakama Nation. The Yakama Nation is located on the eastern side of the Washington Cascades, on approximately 100,000 acres of lands. The Yakama Nation is home to 9,500 tribal members. Much of the reservation lands consist of Douglas-fir, grand fir, and white fir. According to Kelly Olney, Yakama timber manager, the natural fire regime of forest lands are 13 – 20 years. Due to the lack of fire, there has been a 9 ½ billion board feet increase in tree volume, primarily of white fir, since the end of World War II. The tribe's thinning efforts have not been sufficient to keep up with the amount of forest tree growth.

In addition to a large log mill, Yakama Forest Products opened a small log facility in 1997. The two sawmills currently process 143 million board feet annually down from a peak of 225 million board feet in 1999. Yakama Forest Products employs 250 workers, of which 90% are tribal members.

The company is studying the feasibility of building a 16 megawatt biomass co-generation facility at a cost of $25 million. According to Olney, the economics are marginal due to the higher cost of biomass energy.

Plummer Forest Products

Built upon the remains of a big log sawmill that burned to the ground in 1998, Plummer Forest Products is an example of a phoenix rising out of the ashes. Plummer is the only exclusive small log sawmill in Idaho, located in the northern panhandle near Coeur d'Alene. The owner, Todd Brinkmeyer, had the courage to move forward and build a small log facility in 2001. The small log mill currently:

1) Produces 80 million cubic feet of lumber products from trees smaller than 7" in diameter.
2) Processes 45,000 tons of wood chips for the pulp and paper industry
3) Generates 40,000 megawatts of electricity.

The sawmill processes 20,000 logs into 60,000 pieces of lumber every day. According to Brinkmeyer, computer technology was not cheap or fast enough five years ago to work either in the woods or in the sawmills. New software and hardware allows his mill to optimize the small logs into the most valuable products possible.

The sawmill uses 750 truck loads per month. Although national forest lands account for 73% of the forest ownership in Idaho, the mill gets only 5 loads per month from national forestlands. The mill receives less than 30 loads from state lands. Due to the lack of a local small log supply, the company needs to haul logs from as far away as 300 miles.

Much of eastern Oregon and eastern Washington have the potential for setting up of small log facilities as has been developed in northern Idaho and northeast Washington. Savannah Pacific is setting up a small log processing facility in northern Arizona that will manufacture engineered wood products from federal forest biomass.

The big picture of energy production in the US

We're going to shift gears from producing 2 x 4's to producing energy - both electrical and energy to power our vehicles.

We are so accustomed to energy use and consumption in the United States that we don't think about turning a key to start a car or to flip a switch to turn on a light. We don't think for a second that the electrons to power the 60-watt light bulb have been produced from a generator hundreds of miles away.

Energy is produced from both non-renewable resources such as oil, nuclear, and natural gas as well as by renewable resources such as biomass, solar, wind, and hydro. The total amount of all US energy produced is 96.6 quads (Figure 9.2). Don't ask me what a quad is. I'm sure its a pretty big number.

For example, nuclear power produces 8% of the total energy

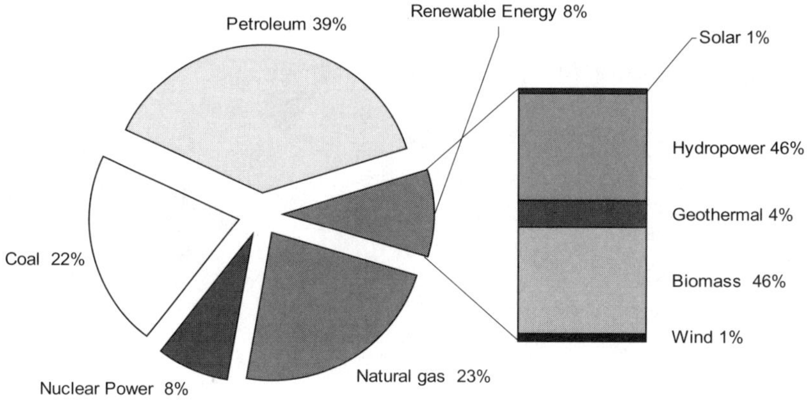

Figure 9-2: US energy production by category
Source: National Renewable Energy Laboratory

consumption in the country. Similarly, renewable energy amounts to 8% of the total energy consumption. The breakout column shows the different types of renewable energy which include solar, hydropower, geothermal, biomass, and wind. The total amount of renewable energy generated is 7.37 quads (Figure 9-4).

Biomass energy production in the US

We're going to be focusing on biomass energy which is produced from wood and/or agricultural byproducts. Go back to the overused "think outside the box" mantra. Even I can understand the concept of taking a tree and cutting the tree into 2 x 4's. You know that mass can be turned into energy (ergo, light this stinking book with a match) or that energy can be turned into mass (plants convert sunlight, water and carbon dioxide into biomass).
But, there are only so many 2 x 6's and 2 x 8's that need to

Figure 9.3 How can biomass energy and products be generated from our forest residues?

Biomass stock
Trees
Forest residues
Agricultural crops
Agricultural residues

Uses

Fuels:
Ethanol
Renewable diesel

Electricity

Heat

Chemicals:
Plastics
Solvents
Pharmaceuticals
Paints, dyes
Detergents

Feed and food

Conversion processes
Gasification
Combustion and cofiring
Pyrolysis
Enzymatic fermentation
Gas / liquid fermentation
Acid hydrolysis / fermentation

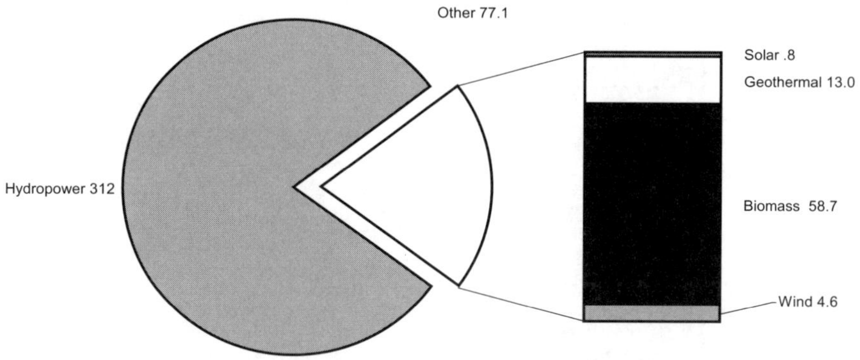

Figure 9-4: US renewable energy production
Source: National Renewable Energy Laboratory

be bought in the United States each year. So, what to do about the billions of tons of woody debris that is sitting, standing, walking, or occasionally flying around that should be removed from our forests.

There are three main types of biopower with another type under development.

1) Direct firing such as the burning of firewood.
2) Cofiring allows the substitution of coal with biomass materials. For example, rather than heating the boiler entirely with coal, the operator mixes in biomass as a blend. The energy efficiency is in the 33% to 37% range.
3) Biomass gasifiers are the third main type of biopower. The biomass material is broken down by heating into a flammable gas composed of hydrogen, carbon dioxide, and methane. The efficiency of biomass gasifiers is calculated to reach 60%.
4) The fourth main type is under development and is known as modular systems.

The amount of biomass electricity generation from 1981 to 1999 has jumped by a four-fold factor in the past twenty years, as displayed in Figure 9-5. There has been an electricity

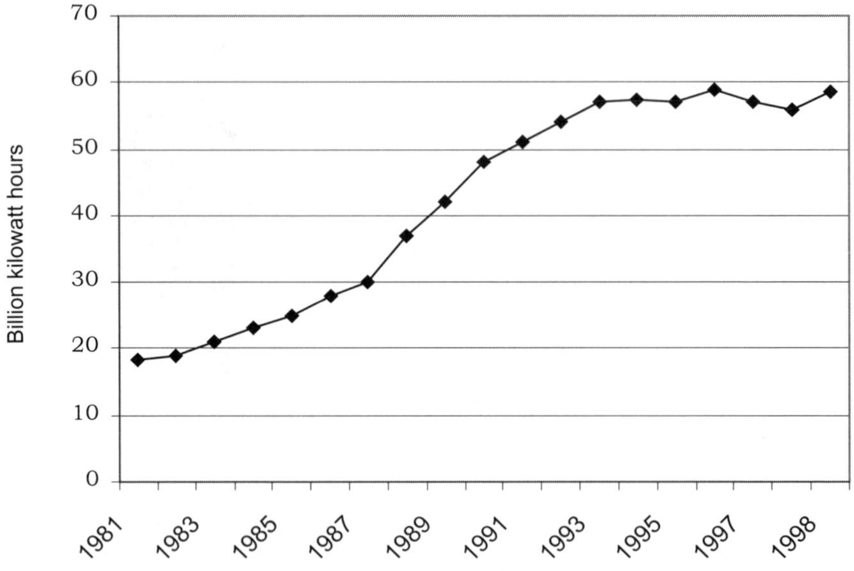

Figure 9-5: Bioenergy electricity generation, 1981 - 1999.

Source: National Renewable Energy Laboratory

generation increase from less than 20 billion kilowatt hours in 1981 to 58.7 billion kilowatt hours in 1999.

You might be wondering why bioenergy electricity generation has leveled off in the 1990s. Two thirds of all biomass electricity is generated by wood products and paper mills. The loss of 800 sawmills in the West in the last ten years has stemmed the increase in biomass electrical production.

Here's the current problem with biomass energy – while its technological feasible, biomass energy can not compete on the open market against other energy sources. Even wind power beats biomass energy hands down. Hydropower currently runs $.03 per kilowatt hour and windpower clocks in at $.05 per kw (Table 9-4).

Most biomass energy is associated with pulp and paper mills or sawmills. Biomass energy is economically feasible mills for two primary reasons – the energy is produced on site and the by-product of energy production – steam – is used by kilns to dry the lumber. So, the co-product helps to subsidize the biomass energy costs.

TSS Consultants and McNeil Technologies developed the following requirements for commercial viability of biomass electrical production.

- Sustainable fuel supply (long-term)
- Fuel costs $10 - $20 per bone dry ton
- Capital costs $1,000 - $1,500 / kilowatt hour
- Power purchase agreements $.10 to $.20 / kilowatt hour
- Identify niche markets

Here's what we can do to create energy self-sufficiency in local communities, reduce fuel loading and generate local jobs. The local community can form a coop, gather up its local resources and buy what is called a BioMax synthesizer. The local community then enters into a long-term stewardship contract with the Forest Service or BLM to remove fuels from the surrounding lands. The power is shifted into the grid which buys the electricity for a ridiculous price of $.18 per kilowatt hour.

The government could calculate how much money that it will save in not having to spend millions of dollars suppressing the wildfires that will occur on those Current Condition 2 and 3 lands in the near future. We know that we need to reduce fuel loading on our public and private lands. We are going to have to spend the money in one form or another form. so we might as well generate electricity.

The US dependence upon imported oil

The United States is becoming more dependent on the importation of foreign oil since the oil crisis twenty years

Table 9-4: Projected biomass cost of electricity				
(cents per kilowatt hour)				
	1990	**2000**	**2010**	**2020**
Utility scale and large distributed power				
Cofiring	NA	2 - 4	1 - 3	1 -2
Direct fired	10 -15	8 - 12	7 -8	6 -7
Gasification	NA	6 - 8	5 -7	4 -6
Small modular - distributed generation				
Solid biomass	NA	15 - 20	8 - 12	6 - 10
Biogas	NA	8 - 12	5 - 8	2 - 8
Source: Biopower Technical Assessment: State of the Industry and Technology, March 2003.				

ago. In 1986, the United States produced 62% of its oil needs (Figure 9-6). Unfortunately, the amount of imported oil from foreign sources has increased substantially in the last twenty years. Today, the United States imports 58% of its oil needs from other countries (Figure 9-7). Many of the oil-producing countries are politically unstable. Oil has a very corrupting impact on developing countries. The OPEC oil cartel can raise its prices as long as the demand holds while the United States can only hope that the price per barrel of oil will not rise too much.

Ethanol offers hope

The three big fossil fuel groups are oil, natural gas, and coal. There are two big problems with fossil fuels.

First, one of the primary reasons that the Earth is warming up is due to the burning of fossil fuels. Upwards of 80% of the United State's energy is supplied by fossil fuels. The burning of fossil fuels emits carbon dioxide into the atmosphere which warms the planet.

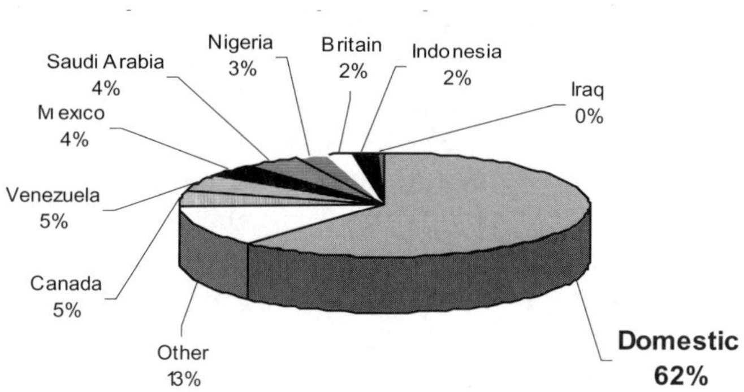

Figure 9-6: Dependence upon imported oil - 1986
Source: US News and World Report

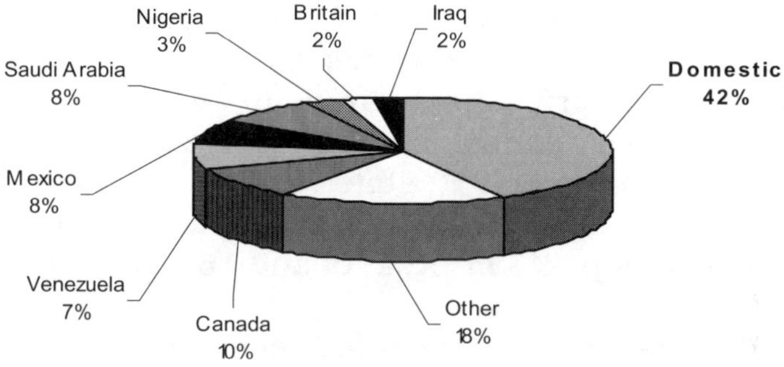

Figure 9-7: Dependence upon imported oil - 2002
Source: US News and World Report

Second, the commercial extraction of fossil fuels will one day no longer be commercially attainable. The amount of oil and gas reserves is debatable, but all agree that the supply is finite. Fossil fuels, by the very nature of their creation, were laid down millions of years ago by carbon-based life forms.

Oil consumption on Earth may very well have passed its half-way point. The *BP Statistical Review* said that in 1989 there were 44.7 years left of consumption. Based upon current rates of consumption, the *Wall Street Journal* estimated that there was 40.6 years of consumption covered by proven reserves in 2002. One of the world's largest oil companies, Royal Dutch/Shell, slashed its proven reserves by 27% in 2004 in a scandal that lead to the expulsion of the top executives at the joint British and Dutch owned corporation. The events of 2004 in the Middle East suggest that oil prices will be likely marching upwards due to the inherent instability of the region.

I have talked about the potential of biomass for electrical production. Another approach to biopower is the distillation of the woody matter into ethanol. If we could figure out how to coax ethanol out of wood, we could tame the 900-lb Middle East gorilla.

You have likely heard about ethanol in terms of agricultural by-products. Ethanol is used as an additive for gasoline. Ethanol has one very big benefit as an additive in gasoline. Ethanol reduces air emissions in the range of 30% to 90%. A number of countries and 13 states already mandate ethanol blending in the range of 5% to 10%. New types of engines allow the use of up to 85% ethanol and 15% gasoline blends.

The United States currently produces approximately 2 ½ billion gallons of ethanol from grains such as corn annually. Brazil, a major producer, generates approximately 3.7 billion gallons of ethanol annually from sugar cane.

The demand for ethanol is expected to increase to anywhere from 132 to 320 billion gallons of ethanol annually by the years 2030 to 2050. This amount of production will require the conversion of approximately 1.43 to 3.16 billion tons of biomass annually. California itself will demand 50% of the US ethanol market in the coming years with its state-mandated clean air rules.

The demand for ethanol is increasing substantially in the United States. From zero ethanol gallons consumed in 2000, Californians consumed 700 million gallons in 2003.

Minnesota consumed 175 million gallons in 2003. What is interesting is that California has no ethanol bioenergy facilities while Minnesota has 14 ethanol bioenergy facilities jointly owned by 5,000 farmers.

Creating ethanol from wood sugars

Dr. Laszlo Paszner, president of Paszner Technologies, Inc. Surrey, British Columbia, Canada has great hopes for ethanol production from wood. I met Dr. Paszner at the small log conference in Coeur d'Alene where he tried to get the attention of foresters and mill owners used to cutting 2 x 4's.

You may have heard of the story of the 300 foresters who fled their home country of Hungary in 1956 with the invasion by the Soviet Union. Dr. Paszner was one of those 300 foresters who were invited by Canada to finish their education and work in Canada.

Dr. Paszner has been trained in both forestry and chemistry. He says that the forest products industry has failed to recognize the true value of what is called wood waste. Dr. Paszner sees great value in wood waste converted into ethanol. He points out that the forest product industry's maximum value of its products is $250 per ton. Dr. Paszner argues that the ethanol production would increase the maximum value to $750 per ton. Dr. Paszner eloquently argues that the industry is failing to get maximum dollar out of its products.

All of the current US ethanol production is coming from agricultural products and waste. Dr. Paszner argues that woody material is more cost-effective than agricultural matter.

Dr. Paszner says that one ton of corn will generate approximately 370 liters (equivalent to 100 gallons) of ethanol. The production cost is $134 Canadian (equivalent to $100.50 US). The direct manufacturing cost per liter ranges from $.32 to $.38 per liter. Direct ethanol manufacturing costs represent 30% of the retail price of gasoline. Translated into our non-metric system, the break even point for ethanol is approximately $2 per gallon.

One ton of wood will yield approximately 370 liters (equivalent to 100 gallons) of ethanol. The direct manufacturing cost per liter ranges from $.07 to $.17 per liter in comparison to the $.32 to $.38 per liter cost using agricultural biomass.

According to Dr. Paszner, the recovery value of softwoods is approximately $2,000 per ton in comparison to approximately $1,300 per ton for sugar cane.

Dr. Paszner is trying to attract venture capital to build a facility in Surrey, British Columbia. Here's the hitch. The investment cost for the plant runs $110 to $145 million dollars. Dr. Paszner claims that there would be a 91% return on investment (ROI).

Ethanol naysayers

Like everyone else in life, ethanol is not without its detractors. Cornell University scientist David Pimentel authored a study that showed that ethanol consumes more energy than it produces and referred to ethanol as "unsustainable, subsidized food burning." Since I did not read the original study, I don't know if Mr. Pimentel had simply examined the energy costs from agricultural wastes or if the study had included naturally-grown wood wastes. In the case of generating ethanol from wood wastes, much of the wood grew naturally without any energy input from our society. I would expect that ethanol generated from naturally grown wood would not consume more energy than had been produced.

On the positive side, ethanol reduces emissions. On the negative side, ethanol gets fewer miles per gallon than gasoline. On the positive side again, ethanol can be readily mixed with gasoline as an additive.

What are the true costs of gasoline consumption?

There are both direct and indirect costs of gasoline consumption. There is the obvious cost of purchasing a gallon of gasoline at the pump. That is not the true cost, however, of the price of gasoline.

Among the many scary things in life is a web site located at www.costofwar.com Similar to the neon sign in New York City that displays the nation's rising debt, the www.costofwar.com has a continuously running counter of the cost of the Iraq war. In August 2004, the cost was estimated at $125 billion and rising at a rate of $1000 every second. The Web visitor was offered other options of how that money could have been spent:

Head Start program	17.7 million kids enrolled
Kid's Health	53.8 million insured
Public education	2.4 million teachers hired
College scholarships	3.2 million 4-year scholarship
Housing	1.8 million units built
World hunger	5 years of fully-funded global anti-hunger efforts
Immunizations	41 years that every child in the world would be immunized.

I would like to add another category. Rather than limit ourselves to a paltry $760 million for 2005 for fuels treatment, we could build 20 ethanol plants at a unit cost of $140 million each for a total of $2.8 billion. Based upon a most efficient level of annual treatment of approximately 5,000,000 acres at $1,000 per acre, the annual cost would be $5 billion for fuels treatment.

In fact, there is probably no way that we could ever possibly figure out how to spend $125 billion on forest biomass reduction and conversion.

Quickie

Given that the Bush Administration is rolling up a $445 billion deficit for FY 04, there is a need to develop business models that will pay for the intensive management work that needs to be performed on our public lands. There are a number of small log facilities that have the capability of handling small logs in diameters down to 5" to 6". A second option is to generate electricity from biomass. A third option - the most promising - is to develop ethanol from wood biomass. Dependence upon foreign oil has risen from 38% to 58% in 2002. Ethanol can be substituted for oil to power cars and trucks and to make the United States less dependent upon foreign oil.

Chapter 10

Global warming and carbon sequestration in forests

Who is this that obscures divine plans
with words of ignorance?
Gird up your loins, now like a man:
I will question you and you tell me the answers!

Where were you when I founded the Earth?
Tell me, if you have understanding.
Who determined its size; do you know?
Who stretched out the measuring line for it?

Into what were its pedestals sunk,
and who laid the cornerstone,
While the morning stars sang in chorus
and all the sons of God shouted for joy?

God speaking to Job
Chapter 38 , Verses 1-7, Holy Bible

Global warming represents a far more serious threat to Americans and the rest of the world's population than any Al-Queda terrorist cell. Global warming does not care if you are an Islamic fundamentalist or a rabid capitalist. Global warming has and will increasingly affect every living creature on this dear planet. Distracted by the fog of the terrorist war, the United States, in particular, is failing to heed the 900 lb. gorilla. That 900 lb. gorilla sitting in our living room is about to reach for the TV remote. While *The Day After Tomorrow* may be sensationalistic in terms of a shortened time scale, we're not going to be able to wrestle that TV remote from the gorilla unless we take action.

In this chapter

This chapter will briefly describe the serious threat of global warming and climate change. The chapter will discuss the potential for the Pacific Northwest's forests to serve as a carbon storage. The chapter will describe how forest landowners and other individuals can make money trading carbon credits on the open market.

The carbon flux is continuous

There has been a continuous interchange of carbon between the Earth and the atmosphere for several billion years. Our human bodies – as well as all carbon-based life forms – are an example of the flow of carbon through the system. Our atmosphere has been greatly modified and changed by the living creatures on this Earth. Most of the oxygen in our atmosphere is the photosynthetic work of plants.

Scientists estimate that 200 billion metric tons of carbon enter and leave the Earth's atmosphere each year. Here's the problem - an estimated 4 to 6 billion metric tons are being added each year. The addition of those metric tons is what is called the "carbon flux" by scientists.

Greenhouse gases have been steadily rising from the pre-industrial ages. Carbon dioxide is the most serious of the four major greenhouse gases. The other three major groups are methane, nitrous oxide, and ozone depleting chemicals. As the percentage of greenhouse gases increase in the atmosphere, the Earth will warm because greenhouse gases prevent the release of radiative heat reflecting off the surface of the Earth.

Greenhouse gases are good in small amounts on the order of 250 – 275 parts per million. Without greenhouse gases, the Earth would be a frigid ice ball like Mars.

Venus

Thick atmosphere containing 96% carbon dioxide. Average temperature is +420 C.

Earth

Moderate atmosphere containing .03% carbon dioxide. Average temperature is +15 C.

Mars

Thin atmosphere. Almost all carbon dioxide is stored in the ground. Average temperature is - 50 C.

Figure 10.1 Carbon dioxide concentrations of Earth, Mars, and Venus

Source: Intergovernmental Panel on Climate Change

Scientists have drilled near Vostok, Antarctica to obtain ice cores that date back for 420,000 years. By measuring the air bubbles in the ice cores, scientists are able to measure the carbon dioxide levels over this period to time. Carbon dioxide levels remained relatively constant for 420,000 years except for the last several hundred years. The parts per million of CO2 during the last 1000 years took an abrupt rise as the 19th century gave way to the 20th century (Figure 10-2). The 2004 level is 370 parts per million. Scientists believe that carbon dioxide levels are higher now than at any period during the past 20 million years.

Strong consensus for global warming

Except for some occasional comments by the Bush Administration or a right-wing magazine about whether the science of global warming has been proven, the consensus of both American and worldwide scientists is

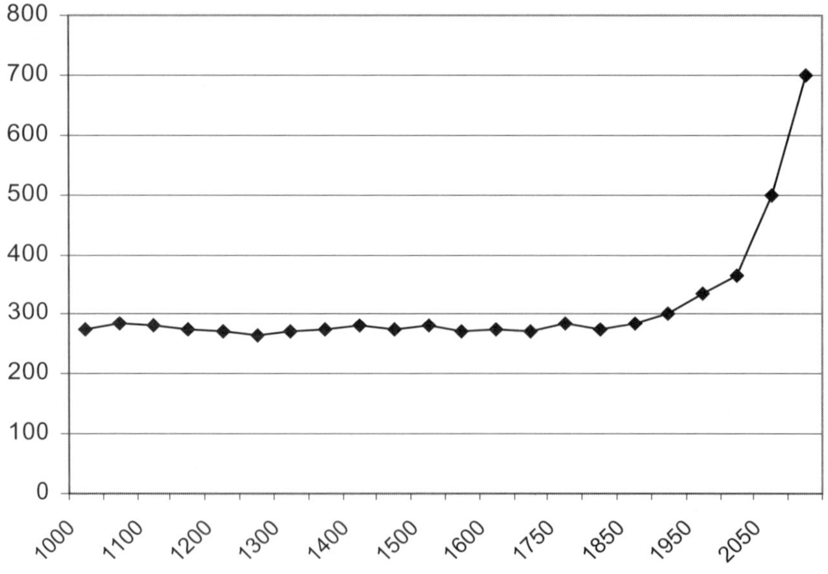

Figure 10-2: Past and projected concentrations of carbon dioxide.

Source: Intergovernmental Panel on Climate Change

that human activities are affecting the atmosphere. The most pre-eminent group studying the problem of global warming is the InterGovernmental Panel on Climate Change (IPCC). The InterGovernmental Panel is chartered through the United Nations and is an outgrowth of the World Meteorological Society.

Climate Change 2001 report

The IPCC has issued three major reports since 1991. The most recent report *Climate Change 2001: The Scientific Basis* issued these findings:

1) The average surface temperature has increased approximately .6 C in the last century.

2) The 1990s was the warmest decade and 1998 the warmest year since reliable temperature measuring instruments were developed in 1861.

3) The increase in Northern Hemisphere temperature in the 20th century is the largest of any century in the last 1000 years.

4) The thickness of the Arctic sea-ice has declined by 40% in recent decades.

5) Warm episodes of the El Nino – Southern Oscillation phenomenon have been much more frequent, persistent and intense since the mid-1970s, compared with the previous 100 years.

6) The concentration of carbon dioxide has increased by 31% since the beginning of the Industrial Era in 1750.

7) Three quarters of the man-made emissions of carbon dioxide in the atmosphere in the last twenty years are attributable to fossil fuel burning. The remaining 25% of carbon dioxide emssions are primarily due to land use clearing in developing countries.

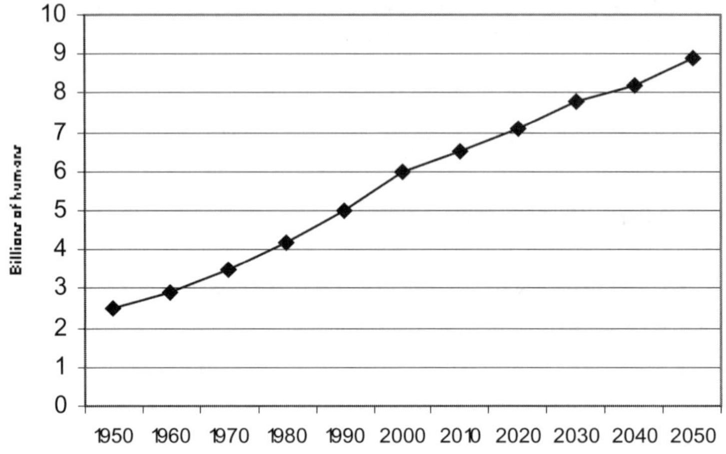

Figure 10-3: Past and projected population
Source: Intergovernmental Panel on Climate Change

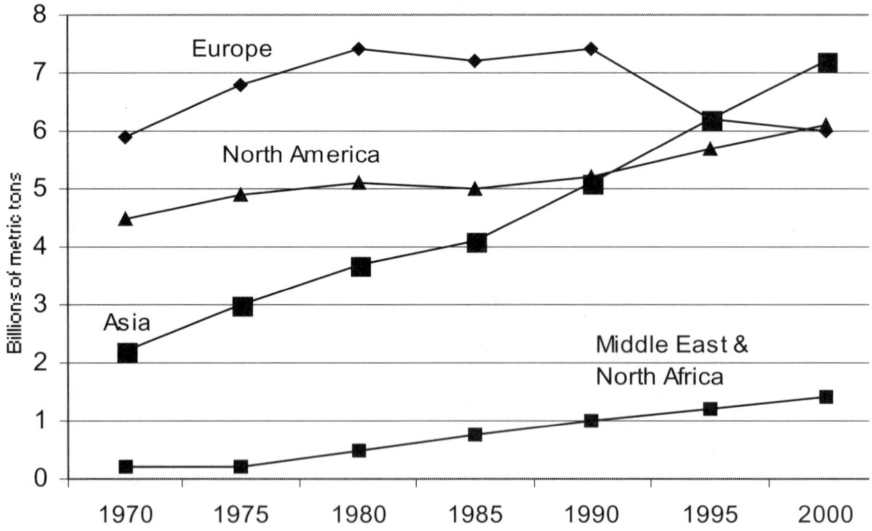

Figure 10-4: Rise in carbon dioxide emissions.
Source: U S News and World Report

Since the report was issued in 2001, the years of 2002 and 2003 became the second and third warmest years recorded in history.

Snow packs have been disappearing in the Pacific Northwest. Philip Mote, Washington's state climatologist and lead author of the regional assessment sponsored by the U.S. Global Climate Research Group, estimates that as much as 60% of the snow pack could disappear by 2050. According to data from 100 snow pack field sites in Oregon, the snow pack dropped from 10" - 15" to slush on Mary's Peak near Corvallis from 1950 - 1997. The 3,750 foot Santiam Junction pass saw an 80% drop while the 4,560 foot Red Butte Mountain saw a 66% loss.

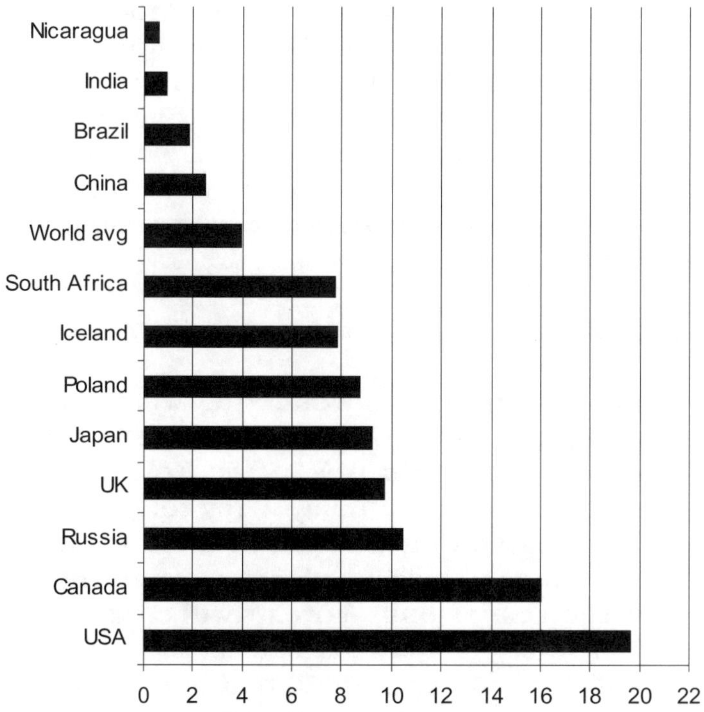

Figure 10-5: Emission tons per capita by country
Source: Intergovernmental Panel on Climate Change

Although the United States has less than 5% of the world's population, this country emits more than 25% of the total carbon dioxide. An average American emits nearly 80 times as much carbon as a Kenyan in Africa (Figure 10.5). However, the United States may soon lose its claim as the worse carbon dioxide emitter in the world as China's and India's economies continue to grow at record rates:

From slightly more than 2 billion metric tons in 1970, the countries of Asia now generate more carbon dioxide than North America (Figure 10.4). To its credit, Europe's total carbon dioxide emissions have declined below 1975 levels because of the concerted efforts to reduce carbon dioxide emissions. In addition, European temperate forests soak up lots of carbon dioxide.

Forest fires release tremendous amounts of carbon dioxide into the atmosphere. University of Maryland fire ecologist Eric Kasischke estimates that the amount of greenhouse gases that are released into the atmosphere during an average fire season in North America is approximately one third of the total exhausts spewed out by all of the world's cars and trucks. As the Earth warms and dries out ecosystems, those ecosystems are more susceptible for wildfires.

Is there any good news about global warming?

Tired of the doom and gloom? Wondering what this has to do with creating jobs in the Pacific Northwest? Here's the good news. Our forests in the Pacific Northwest and throughout the United States can play a key role in helping to capture and pull carbon dioxide out of the atmosphere. Nevertheless, our human race needs to severely put the brakes on our carbon emissions. Forests will certainly play an environmental role, but humans can leverage that role into economics. We should be grateful that our forests have been soaking up huge amounts of excess carbon dioxide.

There are approximately 747 million acres of forests in the United States with an estimated growing stock of 836 billion cubic feet, covering one third of the country (Table 4-2). Growing stock refers to the amount of standing volume of all of the trees in the country.

The total growth is estimated at 30 billion cubic feet each year. Each year, there is an estimated 6 billion cubic feet of tree volume that dies which leaves a net growth of 24 billion cubic feet. Timber harvesting removes about 16 billion cubic feet per year of both live and dead trees.

According to the estimates in *Carbon in US Forests and Wood Products, 1987 – 1997: State by State*, by R. A. Birdsey and G. M. Lewis, US forests gained carbon at an annual rate of 190 million metric tons (Table 10-1).

Table 10-1: Average carbon dioxide change per year

Component	1987	1997	1987 – 97
Biomass	15,833	16,838	100.5
Forest floor	9,401	9,455	5.45
Soils	28,421	28,663	24.1
Wood products and landfills	2,916	3,520	60.8
Total	56,575	58,477	190.1

Source: *Carbon in US Forests and Wood Products, 1987 – 1997: State by State.*

A healthy tree will use up one and a half pounds of carbon dioxide to grow a pound of wood and generate one pound of oxygen. Old growth sequestration rate is less than new growth, but old growth has more stored carbon.

According to the *Analysis of the timber situation in the U.S. 1952 – 2050* by lead technical editor Richard Hayes et al, a one billion board foot increase in timber harvest that is manufactured into structural wood products and used to replace non-renewable products decreases carbon dioxide emissions by 7.5 million tons.

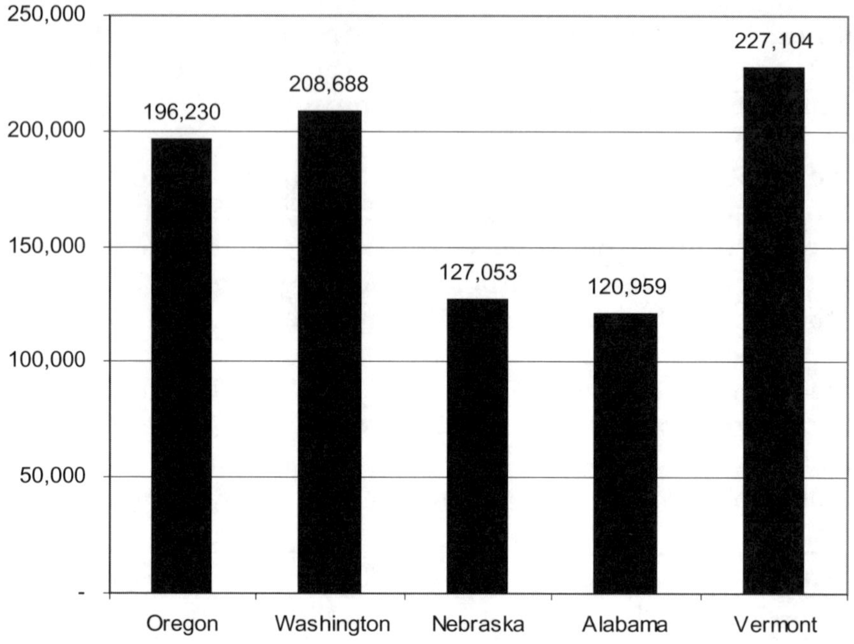

Figure 10-6: Average storage of carbon in lbs per acres
Source: *Carbon in US Forests and Wood Products, 1987 – 1997: State by State.*

Hayes states, "One difficulty is that today we may not recognize an output that may emerge in the future as having value. Two decades ago, few would have identified the ability to store carbon as a valuable ecosystem output."

Carbon storage in the Pacific Northwest

What role does the Pacific Northwest play in the storage of carbon as a carbon sink? At first thought, one would hypothesize that our region stores a significant portion of the carbon. Interestingly, the Northeastern US forests have been absorbing more carbon than our forests in the Pacific Northwest. Some New England forests actually sequester and store more carbon than our Northwest forests (Figure 10-6) on a per acre basis. The difference may be due to differences between old growth vs. new growth and hardwoods vs. softwood composition of the forests.

How much carbon is stored in Oregon and Washington

You might wonder why Oregon is showing a positive amount of 8.9 million tons of carbon storage while Washington is showing a negative carbon flux of 2.1 million tons of carbon (Table 10-2).

Table 10-2: Carbon storage in Oregon and Washington 1987 to 1997 (Millions of tons)				
State	1987	1997	Decade Change	Annual Change
Oregon	2,873.8	2,962.8	89.0	8.9
Washington	2,330.7	2,309.5	-21.2	-2.1
Source: *Carbon in US Forests and Wood Products, 1987 – 1997: State by State.*				

Why is there such a difference? There is a significant difference in land ownerships and management between the two states. Oregon federal lands total more than 50% of the state's forest lands while federal lands amount to only 27% of Washington's forests. However, Washington presumably has more young growth per acre because of a greater percentage of private lands. Young growth will absorb more net carbon than old growth? Or, the difference may be due to poor measurement of the stands.

How to calculate the value of carbon storage

Many ecosystem values don't have a clear, consensus dollar value. For example, how much economic value would you place on a spotted owl? Or, how much economic value would you place on 25 salmon being able to return to a stream because of a retrofitted culvert?

We're not quite so stuck with this problem with the question of carbon storage as markets are slowly beginning to emerge. For example, Oregon State University Professor Mark Harmon estimates the value at anywhere from $5 to $30 per ton. There are a couple of carbon trading markets such as the Chicago Climate Exchange and the COE Exchange. The Chicago Climate Exchange allows the trading of carbon credits. The value in the spring 2004 was less than $1 per ton at the Chicago Climate Exchange board.

Oregon efforts on carbon efforts

The 2001 Oregon state legislature passed a carbon offsets law that establishes a marketable commodity for carbon. The state forester is set up as the broker for this law. You can check Oregon Revised Statues 526, 780-789 for the exact language of the law.

PacifiCorp and Portland General Electric are using trees and forests as a means to offset carbon dioxide emissions. A proposed Klamath cogeneration project is expected to offset 1.16 million metric tons by restoring 2,400 acres of forest over a 100-year period.

Washington efforts on carbon credits

The state of Washington has also been active in carbon credit legislation. The Washington state legislature passed House Bill 3141 in March 2004 that would require energy producers to pay into the fund for each ton of carbon new power plants emit over 30 years. Forestland owners, such as Washington, could receive revenue for each ton of carbon that forests remove from the air as part of a tree's natural respiration.

"With this bill, Washington's state forests can help reduce the amount of carbon in our atmosphere, while receiving revenue for schools and counties," said Commissioner of Public Lands Doug Sutherland. "As state forests grow they remove carbon from the atmosphere, and now energy producers will pay the state to store that carbon. This unique program has the potential to reduce greenhouse gasses and to generate millions of dollars for schools, counties and others who benefit from the stewardship of state forests."

The bill would require power plant operators to offset 20 percent of the carbon dioxide their plants would emit over 30 years at a rate of $1.60 per metric ton, assuming the plant runs 60 percent of the time. That would cost the operator of a new 650 megawatt plant about $11.7 million, which would be used to support an array of environmental purposes. This new bill will provide a boost to the fledgling market for "carbon credits."

Sutherland noted that Washington and other western states are participating in a federal grant to examine how the 2.1 million acres of state trust forestland and 8.5 million acres of private forestland could be used to reduce greenhouse gas that comes from electric power generated by fossil fuels.

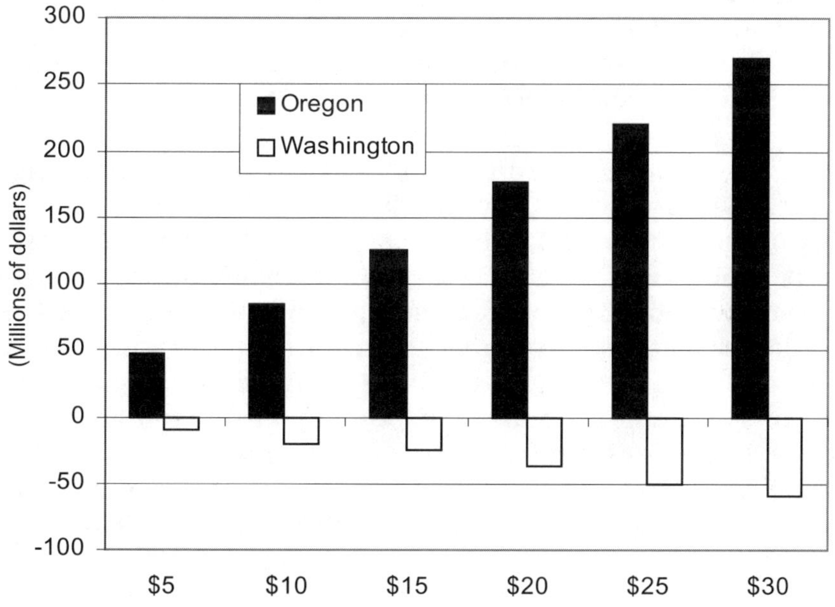

Figure 10-7: Potential carbon storage value in millions of dollars based upon a range of price options

What is the storage value in the Pacific Northwest?

Quoted in the Oregon Board of Forestry's *Forestry Program for Oregon*, OSU professor Mark Harmon says, "The Pacific

Northwest...is a great place to store carbon because the forests here can be long-lived. Those forests are very productive."

Using Harmon's estimates of carbon storage values of $5 to $30 per ton, one could calculate the value (or the cost to residents in the case of Washington state) of carbon storage if there was a fully functioning carbon commodity market (Figure 10-6).

Why isn't there a fully functioning carbon commodity market? The United States deserves most of the credit for stopping this commodity market. You may have heard about the Kyoto Protocol. The Kyoto Protocol was an attempt to get all of the nations together to agree to carbon dioxide reductions. If you heard anything about the Kyoto Protocol, you heard some quick, useless sound bites about how Americans would suffer economically. The US Senate voted 96 – 0 to reject the Kyoto Protocol in 1996.

Americans are recognizing that global warming is becoming a problem. With a grant from the National Science Foundation, Tony Leiserowitz of Decision Research polled Americans about global warming. 86% of Americans supported the U.S. signing the Kyoto Protocol and that 90% said the U.S. should take steps to reduce greenhouse gases. Of course, Americans weren't willing to make many personal sacrifices. 78% of Americans opposed a 60-cent gas tax increase to reduce emissions and driving. All is not lost. The US Senate narrowly rejected a carbon credits trading bill that was called the McCain-Lieberman bill in the fall 2003. It is only a matter of time before the bill is passed.

Quickie

Global warming is the 900-lb gorilla. Forests can play a useful role to capture some of the excess carbon dioxide emitted by human activities. As carbon credit markets develop and mature in the next decade, Washington and Oregon landowners may be able to sell carbon credits on the open market.

Chapter 11

Limitations to job creation

The urbanization of the West created another set of challenges to traditional rural policies. Urban populations and new rural residents often embraced the environmental and social values of rural lands, but viewed traditional economics as harmful of their values and interests. Ironically, those who object to economic uses of rural lands often have consumptive habits that belie their convictions, such as owning large homes and driving big cars. This apparent mental disconnect between production and consumption of natural resources is evidence that the social contract between urban and rural America no longer exits.

Forestry Program for Oregon - 2003
Oregon Board of Forestry

That quotation above is a big mouthful by the Oregon Board of Forestry. I was intrigued that a public board would publicly chastise some of its citizens. The Board of Forestry was probably not too concerned because they realize that few urban people have ever heard of the *Forestry Program for Oregon*. I happen to agree with the Board of Forestry that Oregonians as well as most Americans are much too disconnected from their natural resources. Too much TV watching.

In this chapter

This chapter will discuss both the threats to job creation and the limitations that may be imposed upon job creation. Jobs can only be created in a stable, long-term environment that promotes honesty, trust, and compassion. The forest industry needs to be held financially accountable for their actions while the environmental community needs to be held accountable for their misstatements and their wasteful lawsuits.

Stable and predictable timber supply

I was talking with a logging contractor about his successful stewardship contract for 1,700 acres of commercial thinning on the Winema National Forest in 2002. He remarked that their company had to learn new techniques to perform the stewardship contract. He had worked on the contract for many months to hone their skills and understand how to deliver the product efficiently. Unfortunately, once he finished the project, the Forest Service did not have any more contract work. His biggest complaint was that his cut-to-length unit was idling in his backyard because there weren't any more contracts on the forest.

Environmentalists don't understand the severe impacts for a small business or even a large corporation dealing with an interruptible timber supply. Businesses need to be able to market goods and services in a stable environment that will allow the development of goods and services.

The Oregon Department of Forestry, which owns just 4% of the forestlands in Oregon, had a timber sale harvest of 350 million board feet in 2002. The sale level was more than the combined timber sale harvest on Forest Service and BLM lands in Oregon although the federal government manages over 50% of the forestlands in the state. So, if a logger was to project where he would mostly likely be able to do business in the future, the logger would be quite concerned about attempting to perform work on federal lands.

Non-compliance problems affect the industry

Corruption has a very significant negative effect on economies. If businesses do not feel that there is a "level playing field," honest business people will be disinclined to make the necessary investments to stay in the business or may be inclined to act like the dishonest characters to simply survive in the cutthroat environment.

The logging and forestry services industry have been beset with non-compliant contractors ever since the industry felled its first tree. "Gypo loggers," short on capital funds, try to stretch dollars by creating exotic partnerships so that the loggers are not covered for workers compensation coverage. Forestry service contractors who use undocumented workers illegally are tempted to pay cash under the table to these workers. The workers aren't interested in paying social security taxes since their wages would be reported under a false social security number and they would never get credit for their earnings.

Reforestation association's compliance effort in Oregon

Due to the rapidly rising costs of workers compensation for forestry services contractors in the late 1980s, a group of nine contractors initially formed the Northwest Reforestation Contractors Association (NRCA). The association targeted contractors that it suspected of failing to properly report and pay workers compensation premiums and payroll taxes. The Oregon Department of Insurance and Finance agreed with the association that the 0124 reforestation class code was statistically invalid because of the declining wages during a period when the amount of work was increasing. The insurance department initially ordered a 20% reduction in the pure premium rate for the 0124 class code and ordered that workers compensation carriers such as SAIF Corporation and Liberty Northwest do a better job of auditing contractors.

Since SAIF underwrote over three quarters of the forestry contractors, SAIF proceeded to audit contractors with advisory help from the contractors' association. SAIF nailed one contractor for over $700,000 and another contractor for over $200,000 of unreported premiums. Many other contractors were assessed smaller amounts. NRCA directly sued over 10 contractors for failing to submit certified payroll reports to the Bureau of Labor and Industries. Most of those contractors simply folded their businesses and disappeared.

Association's compliance efforts in Washington

At the request of six Hispanic contractors in Washington, NRCA worked with the Washington Department of Labor and Industries (WDLI) to target contractors who were failing to properly report wages in the early 1990s. An aggressive audit program identified over $4 million of unreported premiums. Although the Washington reforestation rates had been rapidly rising for the past few years, the aggressive auditing program coupled with a requirement that both the landowner and contractor report contract awards significantly increased the amount of payroll reported. WDLI was able to reduce reforestation workers compensation rates by 50% during the audit effort because more payroll and premiums were being reported and paid.

Industry compliance efforts fall off

Fast forward to the late 1990s. A number of contractors had migrated from forestry services to fire fighting services because of the development of an inter-agency fire contract that had been created in the Pacific Northwest. The big fire years of 2000 and 2002 generated lots of work for the fire contractors when the federal government itself paid out $1.5 billion dollars in suppression costs for each year. The Department of Forestry estimated that Oregon fire contractors raked in over $91 million in 2002. Fire fighting, unlike thinning, is very lucrative.

For example, contractors charge approximately $28 per person per hour to the government. A 20-person crew will typically work 12 hours per day on fires or being transported to fires. Multiply 20 people times 12 hours times $28 to yield $6,720 for a daily gross. Contractors typically pay wage and benefits of $8.25 per hour to the majority of the workers. After paying for fire equipment, vehicles, insurance, and workers compensation premiums, payroll taxes, even a dim-witted fire contractor can make 20% - 25% pure profit per crew. And, a contractor can choose how many fire crews to run – one crew, five crews, ten crews, etc.

Barriers to entry were quite low

One of Michael Porter's Five Forces comes into play here – entry costs. The entry costs are quite low – a couple of $20,000 vans, personal protective fire gear, and some training costs.

Like turkey vultures that will flock from tens of miles around, fire contractors and wanna-be fire contractors swarmed over Uncle Sam's carcass. In the mid-1990s, about 50 fire contractors consistently provided a total of 100 crews of 20-person crew workers annually. The National Wildfire Coordinating Group established minimum standards for the crews. These standards included classroom training and field experience requirements for the crew boss and two squad bosses. A reputable contractor will generally need about five years to train a worker to become a crew boss. As word spread about the new "gold rush," existing fire contractors and new fire contractors boosted the number of fire crews. The number of fire crews doubled to 200 crews in 2001; rose to 285 in 2003, and has reputedly hit 365 fire crews in 2004. Training records for crew bosses and squad bosses were falsified as contractors became desperate to find qualified employees.

Quality rapidly declined among the crews. One vehicle wreck in 2002 killed 6 firefighters. A second vehicle wreck in 2003 killed eight more fire fighters. *The Oregonian* ran several front page articles about the problems. The fire associations did little to stop the non-compliance problems except to point the finger at the government agencies.

The problem does not stop there. Contractors have become accustomed to hoping that the forests burn down so that they can make their 20% to 30% profit since they have trouble making a 5% profit on basic forestry contracts. Many contractors subsidize their forestry service contractors during the winter and spring months with the expectation that the forests will burn down in the summer.

Big problems for the special forest products industry

A van carrying 10 pickers and harvesters crossed over the thin, yellow line in the middle of the road near Morton, Washington and smacked headlong into an oncoming truck in March 2004. Five of the harvesters were killed in the vehicle accident. The other harvesters were seriously injured. All of the harvesters were from Guatemala. One of the harvesters killed was 15 years of age. The harvesters were reputedly earning only a couple of dollars an hour colecting special forest products.

Vehicle accidents involving employees can cost millions of dollars in the treatment of medical injuries, lost wages, and permanent or partial disability awards for diminished wage earning. There was no company that accepted responsibility for these dead and injured workers. The sheds which buy from the harvesters claim no responsibility that these are their workers. The agencies and private timber companies that issue harvest leases assume no responsibility for these workers. The owner/driver of the van typically has no insurance (or even the concept of workers compensation coverage for the workers). It's hard to imagine that the workers had personal medical insurance. Who is going to pay the bill for the deaths and injuries of these harvesters and pickers?

Phony environmental issues

Environmental organizations have been extremely slow to understand that the blast furnace fires of today are causing serious environmental impacts. Rather, the environmental organizations have been trying to shift the focus away from the over-stocking problems on public lands. Mindful that the public was becoming concerned about the impacts of fires in the summer of 2003, a number of major environmental organizations in the Pacific Northwest dipped heavily into their lean pockets to run full-page ads in some of the major dailies to encourage readers to draft letters to their elected officials to force the Forest Service and BLM to pay to treat lands closer to urban areas rather than spend the money on distant lands away from the wildland-urban interface.

One wonders if the environmental groups ever bothered to look at land ownership maps in Oregon and Washington. The closest national forest lands in Seattle, Washington are found nearly 75 miles from the city. Other major cities such as Olympia and Tacoma are more than thirty miles as the crow flies from national forestlands.

While a knowledgeable person could easily see through what might be caustically referred to as a disinformation campaign, the urban public would likely not be aware that public lands are so far away. Another question is why should a St. Louis taxpayer be subsidizing a wealthy suburban resident living in a trophy home in the Pacific Northwest woods.

Environmentalists are now the Emperor without clothes

I attended the 30th annual reunion of the Oregon Natural Resources Council (ONRC) in the winter 2004 as I had worked in the early 1980s for the environmental organization to help pass the one million acre Oregon wilderness bill. Andy Kerr, well-known for his pointed barbs at timber industry hacks while working for ONRC in the 1970s and 1980s, spoke at the reunion. Andy took a few minutes to thank his timber industry lobbyist opponents. He joked that those timber industry lobbyists had helped to make him well-known. Andy then talked about future environmental efforts. He began to list off working towards protecting millions of acres of desert wilderness, creating a new national park in southwest Oregon, and protecting several million acres of ocean wilderness. He received only polite applause for the grand agenda.

Elected urban officials go for the easy vote

Even many of our elected officials don't understand the seriousness of the problem and the possibility to create jobs. Former Oregon Governor John Kitzhaber wrote an opinion piece in the February 27, 2004 edition of *The Oregonian* about salvaging timber from the Biscuit Fire. Not surprisingly, the Governor criticized the Forest Service plan and stated, *"Instead of opting for the 'conservative' approach to post-fire salvage called for in the Northwest Forest Plan that guides the management of national forest lands in Oregon and Washington, the administration has chosen a huge salvage effort the likes of which has not been seen in decades."*

I'll preface my comment by noting that I voted for Kitzhaber both times that he ran for the governor's office. *"Excuse me, Governor Kitzhaber, but the Biscuit Fire is the largest fire that has ever burned in the state of Oregon for over a century. Doesn't it make some sense that the 'salvage effort' would be in the 'likes of which has not been seen in decades'."*

Unfortunately, Governor Ted Kulongoski has fallen into the same mind-set as his predecessor. Once again, I voted for Kulongoski and I will vote for him again in two years. However, he has also publicly complained about the Forest Service wanting to harvest too many trees on the Biscuit Fire. It's a cheap urban vote getter for Democrats. Representative Peter DeFazio is the only Oregon Democrat who understands the intricacies of the forest problems.

Representative Les AuCoin (D-OR), who used to represent the First District of Oregon, has become a columnist and appears in my local daily newspaper on a regular basis. He, too, wrote an article complaining about the large size of the Biscuit salvage effort. I had to laugh out loud reading his article. While he was representing the First US Congressional District in northwest Oregon, he sat on the powerful House Appropriations Committee. He routinely held the Forest Service to getting out a sale level of over 4 billion board feet a year in the Pacific Northwest. While working for the Sierra Club and Oregon Wilderness Coalition in the 1980s, I used to meet with his Congressional aides (and even the Congressman himself) to ask that he lighten up on the sale level pressure placed on the federal agencies. What is greatly disturbing is the likelihood that the Kalmiopsis Wilderness and surrounding lands will turn into 500,000 acres of brushlands over the next twenty years. I can't help but contrast the effort of our hillbilly Oregonians who reforested the Tillamook Burns sixty years ago with our failure to recognize the consequences of our lack of action.

Too many so-called environmentalists are knee-jerk reacting to present-day events with 1970s and 1980s thinking about the Forest Service, BLM, and timber industry. While that 1970s and 1980s thinking mentality was largely appropriate for that period of time, the federal agencies and timber industry have changed. More importantly, the living biomass on our lands has changed dramatically. The good news is that Oregonians and Washingtonians are recognizing that a change in thinking is needed.

Trends of public thinking

In order to develop long-term forestry jobs in the Pacific Northwest, the public must be supportive of active natural resource management. There are a number of signs that the public is shifting its attitude to more active management. According to a 2003 Oregon Forest Resources Institute poll, 85% of Oregonians are concerned about the fire danger. The survey found that Oregonians do not believe that forests in Oregon are being managed effectively for fire resilience.

The Washington Department of Natural Resources held a series of public meetings around the state in 2002 to gauge public opinion primarily about timber management. One might accuse the Washington DNR of asking loaded questions since the agency has generated $5.7 billion in timber sale revenue since 1970 for distribution to schools, universities, and the state's general fund. Here's a statewide tabulation of some of the loaded questions asked by DNR:

Value Statement	**1**	**2**	**3**	**4**	**5**	**Value Statement**
I believe timber harvesting can be done responsibly, benefiting the people of Washington by providing one third of school construction costs and still can promote healthy ecosystems	60%	11%	8%	5.5%	15.5%	I believe we cannot simultaneously provide significant funding for school construction through timber harvests and still protect healthy forest ecosystems.
I believe that nature has survived without us, and nature manages better than people; we should be passive stewards of our forests.	4%	8%	14%	17%	56%	I believe interaction between people and nature is inevitable and unavoidable in managing our forests. We should be active stewards of our forests and make forests more healthy and productive.

Okay, this sample of questions are loaded so we can toss out this survey as too self-serving. I myself get pretty annoyed when federal/state natural resource agencies and private timber companies begin to hold themselves out as "stewards" of the land. Many environmentalists blanch at the "stewardship" concept. And, I don't blame them. If you are going to claim that you are a steward of the land, you can't just engage in "command and control."

I am reminded of a Biblical story that I heard one Sunday in our church when I apparently was not sleeping. The story was about a shepherd who led his sheep into a rock holding pen each night. When asked why there was no gate to keep the sheep in at night, the shepherd replied that he slept at the opening in the rock holding pen so that he could protect the sheep from predators.

We hold people liable for their debts that they incur. Similarly, we need to hold stewards liable for their actions.

The forgotten problem: Undocumented workers

While I was working for the Northwest Forestry Contractors Association in the early 1990s, the Emerald Chapter of the Society of American Foresters (SAF) made a serious mistake and invited me to speak at their monthly meeting. Our association had been waging a Don Quixote battle on the illegal alien problem against the big kids on the block.

Our association had been working with the Border Patrol for a number of years to identify the "usual suspects." The Border Patrol would sweep into areas where forestry crews were working. Sometimes, the results were quite dramatic with the capture of over 100 illegal workers. The Border Patrol referred to those operations as "catch and release."

After getting the go-ahead from the Board of Directors, I spoke and identified reforestation crews that had been popped by the Border Patrol on Forest Service, BLM, Weyerhaeuser, and IP lands during the last few months to the SAF members. Needless to say, each of the head foresters from their respective agencies and private timber companies indignantly stood up and declared that they did not knowingly allow the use of illegal aliens.

I briefly mention this situation because much of the intensive management work that needs to be performed on our public and private lands in the Pacific Northwest will be performed by individuals who are not legally entitled to work in our country. For the record, I have since knowingly hired illegal aliens. If you can't beat them, you have to join them.

Running an honest set of books

While a host of CPAs will disagree, many small businesses like to bend the accounting rules to suit their particular needs. Most small businesses are closely held by a few individuals who also sit on the Board of Directors. So, there are not the usual checks and balances that are found in major corporations such as Enron, WorldCom, Tyco, Global Crossing, Qwest, etc. Instead, small businesses are particularly focused on limiting the amount of corporate taxes or personal income taxes that need to be paid.

I was reminded of the lack of transparency when I was a finalist for a controller position at a sawmill. During the first interview, I had mentioned that I run a clean set of books. When I returned several weeks later for a second interview, I was ushered into the existing controller's office to be asked about my opinion of the owners charging off trips to Los Angeles with their families and personal items for their houses on the company's books.

From a strictly accounting position, I gave the wrong answer. I had said that I would judge those expenditures in terms of whether those dollar amounts were material or not. I had said that I would not be concerned with amounts less than $1,000. The controller then complained about the excessively harsh IRS inheritance tax rules that do not allow money to be easily passed down from generation to generation. Although I didn't say anything, there is a one million dollar exclusion on estates so the vast majority of

Americans are easily able to pass their inheritance to the next generation. Needless to say, the company did not hire me. The company was looking for a controller who would willingly allow the owners to charge personal items to the company.

I cite this one small example because small and large businesses need to run honest sets of books. If those companies want to have access to public natural resources, those companies need to pay their fair share of taxes.

Quickie

There are many limitations to sustained employment and job growth. Long-term job growth in this industry requires a stable and predictable timber supply, an honest set of contractors, and a better understanding of the ecological threats facing our forests. The forest industry needs to be held financially accountable for their actions while the environmental community needs to be held accountable for their misstatements and their wasteful lawsuits. Many environmental organizations approach forest management with the same simplistic attitude that the Bush Administration approached the Iraq war.

Chapter 12

How many jobs will be created?

Those who can, work. Those who can't work, teach. Those who can't teach, write.

You've probably been asking yourself about the three p's that a reader looks for in a book: promise, performance, and payoff. You're probably thinking that this Desmond guy has certainly promised quite a lot and that he has thrown a lot of charts and graphs at me to hide his own confusion. You might be wondering if he is ever going to deliver on the payoff. I'll pose the question for you to me.

"So, Mr. Desmond, how many jobs are going to be created? I just want a number. No more charts or graphs. Just a number."

And I say, "21,416 jobs, give or take 10,000 jobs."

And, you then mumble to yourself, "That's it. I'm putting this book down for good."

In this chapter?

The jobs payoff.

How were the job numbers developed?

For those of you who are still reading the book, I'll explain how I came up with this admittedly phony set of numbers of 21,416 jobs, give or take 10,000 jobs.

The forest industry developed a rough rule for the number of jobs that would be created from the harvest of one million board feet of trees. Standard cost accounting refers to this concept as the *direct cost allocation* basis. The rule of thumb is 8 jobs will be created for every one million board feet that is harvested and processed. These jobs include the forester who lays out the timber sale; the logger who cuts the trees; the forestry services worker who plants, sprays and pre-commercial thins the site; the log truck driver who delivers the trees to the mill; the mill worker who processes the wood; and the truck driver who delivers the finished lumber or paper to the distributor or retail store.

We now turn our attention to the number of acres that need to be treated. For purposes of this model, I limited the analysis to Class II (moderate) and Class III (high) national forest lands in Oregon and Washington simply because those raw numbers are available. There are 11.6 million acres of Class II lands and 8.2 million acres of Class III lands in both states (Table 6-4). Chapter 6 included an analysis by Bunn and Hannel that the most efficient level of fuels treatment was approximately 2.7% of the lands needing treatment each year. For ballpark purposes, there are approximately 534,000 acres to treat annually.

The last step is an absolutely wild guess about the amount of board feet that need to be removed per acre. For example, an old growth forest may contain 60,000 to 70,000 board feet per acre in the western Pacific Northwest.

Duane Vaagen is a millowner in eastern Washington who has been harvesting small logs for many years and probably has a pretty good feel for average amounts across a large landscape. Vaagen estimated that there was an approximate average of 5,000 board feet per acre to treat at *Timber West's* Small Log Conference in Coeur d'Alene, Idaho in April 2004.

First, a caveat. In my many years of generating and projecting income and cash flow statements, I can only think of one spreadsheet that I created that was accurate six months later. Some of my spreadsheets missed by pretty impressive spreads.

Here goes the math:

Table 12-1: Number of jobs to be created

Type	Total Acres (million)	Annual treatment	Per acre rate	Total (Thousand Board Feet)
Class II	11.6	.027	5,000	1,569,115
Class III	8.2	.027	5,000	1,108,480
Total				2,677,595

The next step is to calculate the board feet times multiplier effect

2,677,595,000 / 1,000,000 board feet = 2,677

2,677 * 8 jobs per mmbf = 21,416 jobs

The fudge factor of plus or minus ten thousand jobs isn't based upon any logic or previous experience. It's just another number.

What could possibly be wrong with these numbers?

Absolutely everything. This model assumes that the

Forest Service and BLM would have a sufficient workforce and support to ramp up to the ideal levels of treatment.

The American public needs to support active forest management to reduce fuel loadings on our public and private lands. The American public needs to recognize that we should become self-sufficient in energy and wood needs rather than hypocritically import huge amounts from foreign countries. The American public needs to trust professionals working in public agencies and private companies to perform this work for the good of all our society.

Once the American public is supportive of active forest management, the federal, state, and local governments will follow behind to help create those jobs.

The truth is, no one knows what the actual numbers will total. The number of jobs created per million board feet may be higher because of the more labor intensive nature of harvesting much smaller diameter trees. But, mills are becoming ruthlessly efficient and requiring fewer workers. I don't pretend to know how many jobs can be created. I do know that this work needs to get done.

By the way, what is a good job

While the average annual salary for Washington and Oregon workers is $38,244 and $33,684 respectively, that figure is inaccurate as I discussed in Chapter 2. A much more accurate figure is the median wage of Oregon which is just $12 per hour. Given the enormous pressures on wages due to global outsourcing, we need to be pretty realistic about what a good job pays anymore. When 75-100 people will apply for $12 per hour job, then there is strong anecdotal information that there are not many well-paying jobs available. So, my definition of a good job is any job that pays more than Oregon's median hourly rate of $12 per hour.

What will these jobs pay?

Table 2-1 Washington Employment and Wages - 2002 and Table 2-2 Oregon Employment and Wages - 2002 show the average manufacturing jobs are $50,901 and $43,600 respectively.

Table 12-2: Forest and Paper salaries	
Logging equipment operator	$35,868
Faller	$48,576.
Choker setters	$11 - $13/hr
Forest and conservation workers	$20,904
Forest and conservation technicians	$37,694
Foresters	$54,210
Cabinet makers	$26,760
Saw machine operators	$28,130
Woodworkers	$25,974
Paper Goods Machine Setters	$35,372
Source: Oregon Employment Department (OLMIS)	

Remember Porter's Five Forces Model

Our bountiful forests - if noticeably in declining condition as described in Chapter 6 - are unique to the Pacific Northwest, Alaska and British Columbia. The high-value Douglas-fir forests grow only in this region of the world except for some small areas in New Zealand. We can create jobs and industries around utilizing Douglas-fir, ponderosa pine, hemlock, Sitka spruce, and other species that can not be easily grown in other parts of the world (unless we export the raw logs).

The key point about these types of forestry jobs that can be created is that these jobs are somewhat insulated from the ravages of both global outsourcing and global competition. For example, a logger getting paid $15 to $20 per hour thinning forests can not be replaced by a Chinese worker working in a mill in Asia.

While we can import workers (whether legally or illegally) from other countries to perform work on our forest lands, these jobs are created in the United States. These workers will be paid in American dollars and will likely spend much of their money in the local and regional area.

Quickie

21,416 jobs, plus or minus 10,000 jobs.

Chapter 13

Where do we go from here?

To err is human; to repent, divine; to persist, devilish.

Benjamin Franklin

I worked as a business manager for a fire and forestry services contractor for six years. When I first began working at the company, the company was in severe financial straits. I remember one day looking out the window at the rows of fifty vans, six packs, trucks, and water tenders arrayed in the yard. A job applicant would likely look at all of those vehicles, the large vehicle repair shop, the office structure and would likely think the company was very wealthy. I, however, had a much different view since I managed the company's financial books. I calculated that the IRS owned the first row of vehicles. The second row of vehicles was owned by the state Department of Revenue. Both the vehicle shop and the office building were "owned" by the bank. In fact, there was little on the property that the corporation owned free and clear.

We are like Alice walking around in Wonderland. We have allowed our federal government to generate a $380 billion deficit in FY 2003. We are allowing red ink to spill into the future with no present hope that the deficit will ever be reduced. We have allowed the annual trade deficit with other countries to top $500 billion annually.

Americans are not paying attention to the deficits

Our local *Register-Guard* newspaper editorialized about the lack of outrage of Americans towards the federal deficit. The newspaper's editor was inclined to think that maybe Americans were willing to tolerate the huge deficit because they did not understand the big numbers. However, the editor noted that Americans were swimming in a collective consumer debt of $2 trillion dollars. The editor felt that Americans had become so used to debt that the growing federal deficit did not have much impact.

What few Americans know is that the federal deficit and the trade deficit are being financed by foreigners. The Saudis, Chinese, and Japanese collectively pay for over 60% of the federal deficit. When foreigners decide to pull out their money and to invest their money in other markets, the crash will be heard round the world. The U.S. has traditionally been the largest market for new investment capital in the world. However, China overtook the U.S. as the number one country for investments in 2003. Did the U.S. fall to the number 2 position. No, France took the number 2 position in 2003.

What few Americans know is that Argentina was as wealthy of a country in per capita terms at the beginning of the 20th century as the United States. Argentina had abundant natural resources and an educated population immigrating from Europe. Argentina attempted to maintain a high living standard throughout the century, based in part upon borrowed money from foreign sources. The country's finances began to unravel in the late 1990s when the country was no longer able to finance its $140 billion debt. Capital dried up. Argentinean leaders came and went with breath-taking speed. The unemployment rate skyrocketed to 25%. The government shut down the banks and limited the amount of money that could be withdrawn out of banks. The middle class was reduced to selling goods on streets. Riots broke out. Argentineans were quick to blame their government. In reality, Argentineans only had themselves to blame for living in an Alice in Wonderland world.

Few Americans have heard of the Sword of Damocles index managed by one Wall Street investment firm. The Sword of Damocles index measures the ability of a country to meet its foreign debt and properly manage its internal finances. The Sword of Damocles index correctly foretold the collapse of Argentina. There is only one country in the red zone in its index today. I'll let you guess which one country has that dubious distinction. Even the International Monetary Fund has been warning the United States of its fiscal irresponsibility.

We are living on borrowed time

God grants us only so much time on this Earth. Our human bodies are quickly wore down from alcohol abuse, drug abuse, smoking abuse, and overweight abuse. Our economy has been operating on "borrowing" abuse for too long. Like sand running through the hour glass, we've made a pact with the loan shark.

Although Americans work harder than any other people in the world, we're not working smart. We hypocritically import 1/3 of our lumber because we're opposed to cutting of our forests. We hypocritically import 60% of our oil needs because we're both unwilling to allow energy production in our country and we're unwilling to drive smaller, more fuel-efficient vehicles. Like a frog in a pot of gradually warming water, we give little thought to the rising carbon dioxide levels in the atmosphere. Americans are no longer aware that milk comes from a cow, but think that milk comes from Safeway. Wood comes not from the forest, but from Home Depot. Like Frodo and Sam in Lord of the Rings, we've become dependent upon Gollum to show us the way on our journey.

We can choose

We can choose to embrace our forests and to create an economy based upon our unique natural resources of the Pacific Northwest.

We can choose to understand that lumber comes from trees - not from Home Depot.

We can choose to create jobs by investing money in our public and private forestlands to become energy self-sufficient rather than spend hundreds of billions of dollars in the Middle East.

We can choose to produce ethanol from our biomass.

We can choose to create jobs to reduce fuel loading on our public and private forestlands rather than spend one to two billion dollars every year fighting the wildfires.

We can choose to balance our federal government's deficit and our nation's trade deficit.

We can choose to generate electricity from biomass and create more jobs.

We can choose to reduce carbon dioxide emissions into the atmosphere and to provide mechanisms to create carbon trading markets.

We can choose to create a stable and predictable timber supply and generate more jobs.

We can choose to vote in elections.

We can choose to require timber companies and small businesses to properly report and pay their fair share of taxes.

We can choose to chastize and embarass environmental

organizations when they engage in misstatements.

We can choose to hold federal judges accountable for their short-sighted decisions.

We can choose to rebuild the Forest Service and Bureau of Land Management so that the agencies will actively manage and protect our public lands which will create more jobs.

We can choose to reduce our wood consumption and become self-sufficient.

Finally, we can choose to create a future that values the unique abilities and rights of each precious human being on this Earth and our collective responsibility towards each other and our environment. We have that power.

Glossary

Barriers to entry: One of Porter's Five Forces Model that describes forces, such as high capital costs, which limit a company's ability to enter a new market.

Biomass: As used in this book, biomass refers to the trees and brush growing on our lands.

Biscuit Fire: The Biscuit Fire was a 500,000-acre fire that burned in southwest Oregon in the summer of 2002.

Board foot: One board foot measures 12" x 1" x 1". A log truck carries approximately 5,000 board feet.

Buyer power: One of Porter's Five Forces Model that describes forces such as the strength of buyers to force suppliers to provide the goods and services at a lesser cost.

Bureau of Land Management: The BLM manages 262 million acres of mostly desert and rangeland in the United States. The agency manages 2.1 million acres of valuable forestlands in western Oregon.

Carbon sink: Carbon sinks capture and store carbon dioxide. Through the process of photosynthesis, a tree will store one and a half pounds of carbon and release one pound of oxygen.

Carbon source: Sources of emissions of carbon dioxide such as automobiles and trucks. 80% of the energy produced in the United States involves the consumption of non-renewable fossil fuels, which emit carbon dioxide.

Carbon credits: An emerging marketplace that allows the buying and selling of carbon credits.

Clearcut: The cutting and removal of all of the trees on a stand. Clear cutting is generally practiced on private lands in western Oregon and western Washington.

Commercial forestlands: Forest stands capable of growing at least 20 cubic feet per acre per year.

Condition Class 1: The ecosystem is within its historical range and the risk of losing key ecosystem components is low.

Condition Class 2: The ecosystem has been moderately altered from its historical range and the risk of losing key ecosystem components is moderate.

Condition Class 3: The ecosystem has been significantly altered from its historical range and the risk of losing key ecosystem components is high.

Department of the Interior (DOI): Federal land management agencies of the DOI include the Bureau of Land Management, National Park Service, and Fish and Wildlife Service. Indian Nation lands are provided a higher degree of autonomy, but also fall within the jurisdiction of the DOI.

Douglas-fir: The primary commercial tree species of the Pacific Northwest.

Employment: The Oregon Employment Department showed 1.5 million workers generating gross wages of $52.9 billion in 2002. The Washington Employment Security Department shows 2.6 million workers generating gross wages of $101.1 billion in 2002.

Ethanol: Ethanol can be mixed with gasoline to lower exhaust emissions. Ethanol is produced by converting sugars from biomass into alcohol. The U.S. has a long-range possibility to eliminate its dependence on foreign oil imports if the country were to adopt an aggressive strategy to utilize both woody and agricultural biomass.

Even-flow: The practice of limiting harvesting of trees to a level that will never be diminished over a period of time.

Five Forces Model: Michael Porter's model includes buyer power, barriers to entry, supplier power, degree of rivalry, and threat of substitutes.

Forestry and fire services contractors: In Oregon, the 200+ forestry services contractors are licensed by the Bureau of Labor and Industries. In Washington, the 100+ forestry services contractors are licensed by the Department of Labor and Industries.

Forests: Forests cover about 30% and 33% of the world's and of the United States' land masses respectively.

Forestlands: Forestlands include both commercial and non-commercial forestlands.

Forest Service: The Forest Service manages 46% of the forestland base in Oregon and 25% of the forestland base in Washington.

Global job outsourcing: Jobs that are transferred from the United States to be performed in a foreign country with a lower labor cost.

Global warming: The warming of Earth's atmosphere by the accumulation of greenhouse gases.

Good jobs: For purposes of this book, good jobs pay more than the current median hourly wage of $12 per hour for Oregonians.

Hardwoods: Hardwoods are deciduous trees such as alder, dogwood and cottonwoods.

Healthy Forests Restoration Act: This 2003 Act includes provisions to allow categorical exclusions of 1,000 acres for forestry restoration projects, allows limited timber sale harvesting up to 70 acres, and limits appeals to parties involved in the development of the projects.

Industrial forestlands: Those lands where the owner, typically a corporation, owns more than 5,000 acres of forestlands.

K-V funds: Knudtsen-Vandenburg funds that allow the Forest Service to collect funds from timber sales to pay for the costs of reforestation and restorative work.

Monospony: A market where there are many suppliers, but few buyers.

National Fire Plan: A collaborative effort of federal agencies, state agencies, and local communities to address the threat of wildfires after the disastrous 2000 fire season.

National forests: Oregon has 12 national forests and Washington has 6 national forests.

Non-commercial forestlands: Lands not capable of growing at least 20 cubic feet per acre per year.

Natural fire regimes: The historical fire cycle of the ecosystem.

Non-industrial forestlands: Those lands where the owner, typically a single individual or family, owns less than 5,000 acres of forestlands.

North American Industrial Classification System (NAICS): A system used by the government to track and classify different jobs into 21 major categories such as Leisure, Information, and Construction.

Northwest Forest Plan: A plan devised during the Clinton Administration to address the environmental and economical challenges of the Pacific Northwest.

Northwest Forestry Contractors Association: An association of forestry and fire contractors that once had 70 members in the early 1990s.

Oregon Department of Forestry: The state agency manages 779,000 acres of land and enforces the state forest practices act.

Oregon's forests: Of the 61.4 million acres in Oregon, 27.5 million acres are classified as forestlands.

Pacific Northwest: With apologies to Idaho, Montana, and northern California, this book focuses on Oregon and Washington.

Partial cut: A logging practice to selectively remove trees from a stand. Partial cuts tend to be used on drier timber sites.

Personal income: Total personal income includes wages, social security payments, interest, and capital gains. Washington and Oregon residents generated $194 billion and $100 billion in 2002 respectively.

Pulp logs: Logs that are less than 2 ½" in diameter and are primarily used for chips and hog fuel.

Rivalry, degree of: One of Porter's Five Forces Model that describes forces such as the amount of rivalry between competitors in a marketplace. Some degree of rivalry is useful to strengthen the ability of a company to market and grow.

Saw logs: Logs that are greater than 11" in diameter and are of commercial value. Saw logs are typically sawn into 2" x 4s", 2" x 6s", etc.

Service contracts: Contracts awarded by the government to private contractors to perform services such as tree planting, tree thinning, and fuels reduction work.

Site preparation: The treatment of lands after a logging operation.

Small logs: Logs that are greater than 4 ½" and less than 11 ½" in diameter. Some small logs can be converted into sawn lumber while other small logs are chipped for hog fuel.

Softwoods: Softwoods are typically evergreen trees. Examples of softwoods include Douglas-fir, western hemlock and ponderosa pine trees.

Stewardship contracting: A new form of government contracting that allows for the trading of services for goods. For example, a contractor will thin out a stand of trees in exchange for obtaining the thinned trees.

Supplier power: One of Porter's Five Forces Model that describes forces such as the ability of suppliers to increase prices and control the market because of limited supplies.

Threat of substitutes: One of Porter's Five Forces Model that describes forces such as the ability of a buyer to obtain a different source of goods. For example, a buyer may choose to build a structure with wood, steel, concrete, or bricks.

Timber sales: The contractor pays the government for the harvest and removal of the timber.

Unemployment rate: Oregon and Washington unemployment rates have typically tracked 1.7% to 1.9% points higher than the rest of the nation in the last three years.

Washington's forests: Of the 42.4 million acres in the state, 21.8 million acres are classified as forestlands.

Washington Department of Natural Resources: The state agency manages 2.1 million acres of stand lands and generates $300 million annually – primarily through its timber sale program.

Wildfires: Wildfires have been trending upwards from an average of 2 ½ million acres nationally in 1992 to over 6 million acres in 2002.

Wildland urban interface zone: The mixed land use zone, which includes homes, other buildings and forests or rangelands. The wildland urban interface zone faces challenges to firefighters because of the potentially high costs to defend individual homes from wildfires.

Bibliography

Analysis of the Timber Situation in the U.S. 1952-2050. Haynes, Richard W., Technical Coordinator; General Technical Report PNW-GTR-560, 2000 USDA Forest Service RPA Assessment, http://www.fs.fed.us/pnw/pubs/gtr560/.

Assessment of local thinning and restoration needs, capacity and utilization: Gaps and Opportunities in the Siuslaw Basin. Spencer, Charles; Ecosystem Workforce Program, University of Oregon. February 2004.

Bible, New American Bible, Oxford University Press, Inc. 1990.

Business and Employment Effects of the National Fire Plan in Oregon and Washington in 2001. Moseley, Cassandra; Toth, Nancy; Cambier, Abe. Ecosystem Workforce Program, University of Oregon. December 2002.

Carbon in US Forests and Wood Products 1987 - 1997. Birdsey, R.A.; Lewis, G.M. Gen. Tech. Rep. NE-310. Newtown Square, PA: U.S. Department of Agriculture, Forest Service, Northeastern Research Station.

Case of the Missing Carbon. Appenzeller, Tim. National Geographic, February 2004.

Climate Change 2001: The Scientific Basis. Watson, R.T. and the Core Writing Team (Eds.), Intergovernmental Panel on Climate Change, Geneva, Switzerland. pp 184.

Collaborative approach for reducing wildfire risks to communities and the environment. National Fire Plan, http://www.fireplan.gov/.

Competitive Strategy: Techniques for Analyzing Industries and Competitors. Porter, Michael; Free Press; ISBN: 0684841487.

Covered Employment & Wages, Oregon Employment Department, http://www.qualityinfo.org/olmisj/OlmisZine.

Current and potential energy products from Biomass, Bain, Richard L., National Renewable Energy Laboratory, Golden, Colorado. Powerpoint presentation given at the Healthy Landscapes, Thriving Communities conference, January 21, 2004.

Development of coarse-scale spatial data for wildland fire and fuel management. Schmidt, Kirsten; Menakis, James P.; Hardy, Colin C.; Hann, Wendall J,: Bunnell, David L. 2002, General Tech. Rep RMRS-GTR-87, Forest Service, Rocky Mountain Research Station.

Earth's oil gauge on low?. Huntington, Chris; http://money.cnn.com/2004/05/20/market/oil_reserves/, May 20, 2004.

Employment Transitions in Oregon's Wood Products Sector. Adams, Darius M.; Helvoight, Ted; Ayre, Art; Journal of Forestry, June 03.

Federal agencies announce guidelines for stewardship contracts and agreements, January 15, 2004. www.doi.gov/news/040115a.htm.

Fire and land management planning and implementation across multiple scales. Wendall J,: Bunnell, David L., International Journal of Wildland Fire, 2001.

Forests and Fish Forever, Washington Forest Protection Agency, www.forestsandfish.com.

Forestry Program for Oregon - 2003. Oregon Board of Forestry, http://www.oregonforestry.org/fpfo/2003/.

Future of Energy Policy. Wirth, Timothy; Gray, C. Boyden; Podesta, John D.; Foreign Affairs, July/August 2003.

Future of Earth Special Edition. Duffy, Brian; U. S. News and World Report. March 2004.

Healthy Forests Initiative news release, Release No. fs0405.03, www.usda.gov/news/releases/2003/12/fs0405.htm.

Healthy Landscapes, Thriving Communities, Bio-Energy and Wood Products Conference. Sponsored by the US Departments of Agriculture, Interior and Energy, the National Association of Counties, and the Western Governors' Association, January 2004, http://www.nifc.gov/biomass_conf.html.

Inventory of U.S. Greenhouse Gas Emissions and Sinks: 1990 -2001, U.S. Environmental Protection Agency, April 15, 2003.

Land Commissioner Doug Sutherland praises new law to reduce greenhouse gasses. News release dated March 31, 2004. Washington Department of Natural Resources.

New jobs are for lower pay, study says, Keil, Richard; Arnold, Laurence. Bloomberg News, April 13, 2004.

Private Timber Harvest Potential in Eastern Oregon, Adams, Darius M.; Latta, Greg; Unpublished draft, May 15, 2002.

Renewable Resources Handbook, Chapter 60 - Stewardship Contracting, US Forest Service, http://www.fs.fed.us/forestmanagement/projects/stewardship/handbook/index.shtml

Small Log Conference on Utilization. March 31 - April 2, 2004, Timber West Magazine, Coeur d'Alene, Idaho; www.forestnet.com.

Stewardship Contract Pilots: Monitoring / Evaluation Reports, Kaufman, Marcus. Ecosystem Workforce Project, University of Oregon, 2003.

Technology in the Mills, Eco-Link, Temperate Forest Founcation, Volume 13, No. 2. 2003.

Timber Harvest Projections for Private Land in Western Oregon. Adams, Darius M.; Schillinger, Randall R; Latta, Greg; Van Nalts, Andrienne., College of Forestry, Oregon State University, May 2002.

Truth about America's Forests, Peterson, Jim, Editor; Evergreen Magazine, Spring 2003, www.evergreenmagazine.com.

U.S. Climate Action Report 2002, US Department of State, May 2002, http://yosemite.epa.gov/oar/globalwarming.nsf/content/ResourceCenterPublicationsUSClimateActionReport.html.

Washington State Labor Market and Economic Report - 2003, Employment Security Department, http://www.wa.gov/esd/lmea/pubs/pubs.htm, 2003.

Index

D

E

F

J

K

L

M

N

O

P

R

S